The life so short, the craft so long to learn

GEOFFREY CHAUCER *The parlement of foules*

Contents

List of figures

Preface

THERE ARE MANY people to thank for assistance directly, or indirectly, throughout the conception, gestation and writing of this book.

The idea was born long ago in Aberystwyth when I first became interested in the problems of managing change. A period of sabbatical leave from the University of Cape Town gave me an opportunity to think and to start work; the growing impatience of my family encouraged me to finish it. Colleagues provided great insight when they described their own experiences of change. Helen Carley and the editorial team of Library Association Publishing coaxed, cajoled and provided much-needed critical comment via electronic mail.

In essence, this is a partner to my first book, *Managing change in libraries and information services: a systems approach* (London, Bingley, 1990). In that I described some of what I would now call the 'hard' systems techniques. Both books can be read quite independently of each other, however, and there is little overlap between them.

Peter G. Underwood
Cape Town

Chapter 1

Thinking, analysis and systems

SOFT SYSTEMS ANALYSIS is one of the methods which can be used to analyse the needs of individuals and groups of people who are working in, or affected by, what goes on in an organization. This type of analysis can be of value to those responsible for the management of an organization, since it can offer insights into the behaviour of the people concerned and can also assist in developing and implementing solutions to problems. It is not a method which is confined to analysing needs within the context of libraries, information services and resource centres and it did not originate within that context; it is, however, a method which offers some advantages when used in this particular frame of reference.

Although soft systems analysis was developed in the early 1970s, and has been used successfully for gaining a better understanding of a wide variety of problems, there are few published examples of its use concerning libraries, information services and resource centres. Much of this book is based upon my own experience and the problems I have encountered when using it within this context during periods of consultancy; inevitably, this makes what follows a personal account and I have chosen to use the first person when referring to my own experience, in order to allow you to distinguish between items and ideas which are part of a generally accepted corpus of knowledge and those which are derived from my own experience.

Discussing work based on consultancy experience imposes an additional constraint: that of confidentiality. Where I have made use of such experience I have changed names and disguised circumstances in order to protect that confidentiality. Other than this, details are recorded faithfully to enable you to see how soft systems analysis was used. Bear in mind, though, that you are seeing those incidents through my eyes and mind; had you been present you might have perceived other aspects and seen some aspects differently. Using soft systems analysis does not necessarily mean that all the participants will agree but it does encourage them to reveal and discuss their differences.

What am I trying to do?

Discovering what soft systems analysis is about is best achieved by doing it. Some knowledge of the theory underlying this type of analysis is an obvious prerequisite but deep understanding of its purpose comes from discovering a way of perceiving and working with people, and the situations in which they find themselves, rather than from metaphysical debate.

In writing this book I have kept in mind four objectives:

- to introduce and explain the use of soft systems analysis as a means of exploring the potential for change in organizations
- to discuss how soft systems analysis can be applied in libraries, information services and resource centres
- to provide some examples, mostly drawn from personal experience, of how soft systems analysis can be used within this context
- to assess the relevance of soft systems approaches for problem analysis.

Reading about soft systems approaches, or other techniques of problem analysis, will not turn you into a practitioner any more than reading about management will necessarily make you an expert manager. The theory provides a framework for understanding: experience of use fills the framework with knowledge about what works and does not work in a particular context. My purpose is to persuade you to try soft systems analysis and reflect on the results.

Intended audience

Contact with people, problems and opportunities is a large part of life for most of us, but in the context of work it is middle management and senior management that carry most of the responsibility for dealing with them. In particular, this level of management can make decisions which have a profound and long-lasting effect on the allocation of resources, thus affecting the ability of people to solve problems and make good use of opportunities in the workplace. Soft systems analysis is an appropriate way of considering such decisions: this book should be of interest to anyone who is expected to carry such responsibility.

The use of soft systems analysis within a framework of consultancy is often mentioned in this book. This is because most of my experience of its use has come from such work; it is not my intention to suggest that soft systems analysis should only be used by consultants or within an episode of consultancy. Consultancy is not about prescribing solutions: rather, it is about trying to assist a client to analyse, explore and understand a problem or opportunity and the consequences of various actions which might be taken. In this sense of facilitating deeper understanding, effective managers frequently use the skills of consultancy as a natural part of their management style. Whilst the ideas presented may be of interest to consultants, the book is not primarily addressed to them but rather to those people in libraries, information services and resource centres who have a significant responsibility for coordinating and directing the work of others.

To those who are interested in the work of consultancy, I should like to add a warning. No book can teach anyone how to become a consultant because no book of this kind is ever complete: the practice of consultancy exposes the consultant and the client to a fresh range of views and conditions on every occasion. Be prepared to take the ideas presented and, if they seem to have merit, try them and develop your own style, guided by the experience of others.

What this book contains

This first chapter is about setting the context, explaining some of the terms you will encounter and something of the history of the development of soft systems analysis. In the next two chapters I have discussed planning and what is entailed in working with a group. This is because soft systems analysis is most often used when planning or seeking to respond positively to some change and because the discussion of such planning or response almost always involves other people.

Chapter 4 provides an overview of soft systems analysis, with the intention of identifying its components, its activities, and the sequence of their use. It is intended to be used with the next two chapters, which discuss the two principal tools of soft systems analysis: interviews and models. Taken together, Chapters 4, 5 and 6 are the core of the book and, if you are confident about your understanding of planning and how to work with groups, reading of the preliminary chapters may be unnecessary.

The final chapter provides a bridge between the 'systems world' explored in the core chapters and the 'real world' which we inhabit as managers, consultants, workers and people going about our lives. Its purpose is to explain how to translate the ideas gained from a period of soft systems analysis into a set of requirements which resource managers can put into action.

Terms used in the text

Throughout this book the phrase 'libraries, information services and resource centres' is used. This is a conscious choice, reflecting my understanding of the fact that whilst each focuses on work with information, each is also rather different in the nature of its organization, purposes and clientele. Libraries have, traditionally, emphasized ownership of collections and users have come to them to make use of the stock; even though access to information from external sources is now being facilitated by developments in information technology I still think of a collection as being the principal feature of a library. Information services have, by contrast, emphasized the provision of information, often re-packaging it to suit the individual needs of their users. Resource centres I see as providing means to use information, either through access to sources or ownership of materials, and through access to people who can assist in using this information to best advantage in projects and other similar work.

There are obvious dangers in attempting to make such definitions and distinctions: not everyone will agree with them and they are not necessarily reflected in the names given to the various organizations involved. There are many public libraries which provide excellent information services and there are resource centres which have more than adequate collections. In the course of time the development of information technology may blur the distinctions just made even further until we reach a point when all that can safely be said is that there are organizations whose business it is to facilitate access to information and its use. My aim is to be wide-ranging; hence the use of an inclusive phrase.

A similar difficulty attends the choice of name to be used to describe the people who work within libraries, information services and resource centres. I have chosen to call them 'information workers' and have made no distinction between those with professional library and information science qualifications and those without. The reason is that the distinction is unimportant for the topic of this book; it is also a recognition of the fact that the 'professional versus non-professional' divide does not necessarily characterize all libraries worldwide and is certainly not a major feature in information services or resource centres. There are many examples of people without library and information science qualifications working with information and users in a thoroughly competent manner; this book may well contain ideas of value to them as well as those who are professionally qualified. Once again, I have tried to be inclusive.

A distinction which can be made, and one which I need to make, is between managers and those who are putting into action the decisions of managers. For want of a better term I have called the latter 'staff' but this choice of terms should not be taken to imply that one is automatically of more importance than the other. If being a 'manager' is about

- development of policy
- choice of strategies
- facilitating change

- control of the organization
- use of resources.

then being a 'member of staff' means putting the results of strategic choice into action, selecting and making use of appropriate resources, reporting to a manager on progress and participating in change. Whilst it is common to find a formal hierarchy in an organization, with some people having been assigned the role of manager as a full-time activity, there are other examples where this role is shared amongst a group and allocated to the person having the most experience relevant to the problem or task in hand. Soft systems analysis will be of use with either method of organization but it is often helpful to acknowledge the distinction between 'management' and 'staff' roles so that a clear responsibility for taking action following the analysis emerges.

'Users' is the term I have chosen for those who, in some way, turn to the services offered by information workers and libraries, information services and resource centres. The term 'readers' suggests particular kinds of activity which may or may not be characteristic of all users; other terms, such as 'consumers', suggest a one-way relationship (we provide, they consume) which should not represent how most information workers perceive their jobs.

Another term which I have used is 'stakeholder'. This is a convenient collective word to describe anyone who has an interest in the activities and survival of an organization. For many libraries, information services and resource centres the stakeholders could include users, staff, managers, suppliers, a local community, a parent organization, professional associations, trade unions and government departments. 'Stakeholders' can include individuals, small or large groups of people and entire organizations. Soft systems analysis tries to reveal the stakeholders and what their distinctive interests might be.

The word 'client' has been used to describe someone who employs a consultant. The consultant is answerable to the client (who will usually be a senior manager), may be from outside the organization and may be paid. The relationship is contractual.

'Problems and opportunities' is a phrase used to describe the area of common interest to clients, consultants and participants in the process of analysis. I have favoured this phrase, rather than using the term 'project', because it serves to emphasize the factors which trigger analysis and also reminds us that the two are opposite sides of the same coin: a problem can easily become an opportunity and *vice versa*. It all depends on your perspective.

What is 'analysis'?

It is important to recognize that all methods of needs assessment and analysis have their strengths and weaknesses and that each has a proper use: just as a toolbox contains the means to do useful and beautiful work, so methods of analysis and needs assessment can produce elegant and satisfactory statements if used correctly. Choice of an inappropriate tool or method or its misuse can also have displeasing or disastrous results. This book aims not only to describe soft systems analysis, but how and when to use it.

Before beginning to describe the method and its use it is worth considering what is meant by the term 'analysis' and placing it within the context of libraries, information services and resource centres. By this means a clearer understanding of the scope of what can be done will emerge and some of the general problems associated with analytic procedures will also become apparent.

A potent part of our survival mechanism is an ability to organize what we experience through our senses: we are able to perceive events which impinge upon us and, if we have encountered similar events before, assign them to membership of a group of such events. A further step we take is to discover that such grouping also helps us in deciding what to do about the event: if a particular behaviour in relation to the event seems to have been beneficial in the past, then we are likely to repeat it if the event is encountered again.

There are several dangers attendant on this behaviour, of which two are of special significance. Firstly, there is the problem which might arise if the chosen grouping is wrong: in other words, if the event is not actually of the type it has been perceived to be. Any actions taken may then prove to be incorrect, with consequences which are unexpected or even dangerous. Another problem may be that the surroundings, or the context, of the event has changed but our behavioural response to the event has become instinctive; in other words, the response is no longer appropriate.

To counter either danger it is necessary to try to understand the event: what triggers it, what mechanisms or processes are involved and what effect it has upon its surroundings, especially for the participants. A greater understanding of these aspects and how they interact can lead to *prediction* and, in some circumstances, the ability to *control* the outcome of the event. Analysis is a major part of the process of seeking this understanding and of developing an *explanation* of the event.

In the course of developing understanding there may be events which we observe always to follow the same pattern: they are not subject to variance and, provided the context remains exactly the same, the event will be repeated. This is the nature of a 'deterministic' event: it is invariant and thus predictable. Scientific investigation is driven largely by the attempt to understand events, to seek causes and propose explanations which, if they can be shown to be invariant, can then be expressed as 'laws of nature'. The methods of analysis used in scientific work depend on being able to control, filter or exclude factors which might have an influence on the event but which are not regarded as being principally responsible for it. The scientific method tends, therefore, towards simplification by trying to discover the main factors which determine the course of an event. In order to do this, scientists use the controls and filters mentioned above to lessen the complexity of what is being studied and to build a 'strong' model in which the interaction of a few factors can be studied, analysed and described. In this way, for example, the current in an electrical circuit can be directly related to two factors: the applied voltage and the resistance of the circuit.

Although scientific approaches to explanation have been very successful in building an understanding of our surroundings, there are many events which do not appear to be capable of explanation in a deterministic way. The behaviour of the climate, for example, cannot be predicted with complete accuracy although it is possible to study it over a sufficient period of time to notice certain events which happen with greater regularity than others. The movement of ocean currents may be associated with patterns of rainfall, but it is not possible to predict with complete precision how much rain there will be and where and when it will fall. The understanding that is gained from this type of observation is that, if the ocean currents are observed to move in a particular way, there is a strong likelihood of a particular pattern of rainfall also occurring. What has been developed is a 'probabilistic' model which shows how likely the occurrence of this event will be. It is not, however, as strong as a deterministic model because the outcome is not invariant: other factors which may not be completely recognized or understood have an unpredictable effect on the event. Nevertheless, the model may still be strong enough correctly to predict the outcome in a high proportion of cases, and this adds to understanding of the event. The model can be improved (made 'stronger') by teasing out the effect of the additional factors and investigating the departures from the expected outcome so that the contribution of the additional factors may be understood. The probabilistic nature of the model is reduced and gradually replaced by greater determinism.

There are many limitations to this process of refinement, attributable to both a technical inability to control factors (how could one control all factors in order to predict the weather pattern of the world?) and to problems of subjectivity influencing the interpretation of observations (experimenters may disagree on results obtained from the apparently objective action of reading instruments). Despite improved technical understanding and instrumentation there will remain areas of our experience for which we have little or no explanation or for which our explanations must remain tentative.

Models are generally strongest in the physical sciences, because it is often easier to set up experiments which allow for the control of factors just described and because what is being used for the experiment can be kept largely independent of the observer. The essence of the social sciences, however, is the study of people in order to find explanations of behaviour. Here the problem of control is much greater, especially if one is dealing with large groups of people. There is also the problem of a human being observing other human beings and trying to deduce reasons for particular events: the presence of an observer may influence those events. The 'self-consciousness' of an observer will affect what is recorded and how it is interpreted. Ethical problems also abound and interact with experimental procedures: to what extent, if any, is it morally correct to conduct experiments on people who have not been asked for their consent; to what extent will knowledge of the experiment, and agreeing to take part, influence the behaviour of participants? Such issues are frequently the subject of debate which itself has an effect on experimental procedures. There are practical and moral limits, then, on the procedures which can be used to study society.

These limits affect the methods of analysis used. In the physical sciences an investigation often starts with an observation which leads to a tentative statement, or 'hypothesis'. This, taken together with the laws already accepted as being associated with the event, may lead to an explanation of the event. To become more than tentative, an hypothesis must be tested by constructing and conducting an experiment under controlled conditions.

In the social sciences, an observation may also be the starting point of an investigation; thinking about the possible link between cause and event may also result in the development of an hypothesis. For example, in observing a queue for the tills in a supermarket we might observe that most customers display signs of impatience if kept waiting for more than five minutes; an hypothesis would be that customers will have a more favourable view of the supermarket if queuing time is kept below this threshold. In the absence of laws, however, how can such an hypothesis be tested? The common-sense answer is that one does the next best thing: test the hypothesis against the body of experience of similar events and of other attempts at explanation ('theories') in order to check to what extent it complements or contradicts them. A deeper consideration of this approach reveals that it is also fraught with problems of the subjectivity of the observer and of definition: what is meant by 'a more favourable view'; how can we recognize 'impatience'?

Such a conclusion is quite discouraging and may even persuade readers that there is little to be gained from this type of investigation. Coupling the general problem of observing, of developing and testing theories of behaviour with the recognition that the study of libraries, information services and resource centres is principally concerned with the behaviour of people in relation to the use of information, the person wanting to gain some insight may consider that there is little that can be done to study that behaviour and to ascertain needs within a rigorous paradigm.

Help is at hand, however, if the notion of trying to develop or ascertain generally applicable laws or universal models is abandoned for something more modest. If the focus shifts from the large scale to the smaller scale of seeking to understand the behaviour, problems and needs of particular groupings of individuals and organizations, then useful insights can be gained which will be of immediate benefit to those individuals and organizations. In turn there is the longer-term hope that the accumulation of such insights will eventually lead to the development or refinement of theories of behaviour.

Why raise this topic at all? In writing any textbook about aspects of the management of libraries, information services and resource centres there is a temptation to plunge in, to 'get on with the job', and describe techniques and procedures. Whilst this may be defensible when dealing with something of an abstract and conceptual nature, such as the development of a new system of classification or a cataloguing code, the same cannot be said of describing aspects of managing people. The process of analysis will have an effect on any group of people associated with its use, either as participants or observers. The objective of the analysis is to provide better understanding in advance of some kind of intervention. A sense of responsibility must pervade all such interventions in relation to analysis: part of this is an understanding of the role of analysis and its limitations.

The principal role of analysis in an organization is to provide managers with better insights into the needs of small groups and individuals. The limitation is that these insights are provisional and temporary: in the absence of theories and laws of behaviour there can never be any certainty that the insights are correct and complete or that they are either universal or eternal. This means that insights gained whilst working with one group cannot be applied to another group without first checking that the circumstances are sufficiently similar to make this feasible. Even if there is a superficial resemblance and apparent similarity of circumstances, a deeper study will almost certainly reveal factors which are different. Such factors are likely to affect outcomes in the short or long term in ways which may have a significant bearing on the problem being explored.

The role of the analyst is to assist management to develop models which can aid understanding of a situation, a problem or the needs of individuals and groups. As previously explained, such models cannot be 'strong' and they can only be provisional; they can, if used intelligently and with due caution, assist in the development of greater insight and forestall the making of inappropriate decisions. At the same time, both the analyst and the manager must recognize that the process of analysis is a craft[1] rather than an exact science and this must limit its predictive ability. Knowledge of a craft and its associated skills implies the ability to make the best use of what is available and produce useful results from such material.

Structural matters

When Adam Smith discussed the division of labour in the opening chapter of *The wealth of nations* he used the example of the trade of pin-making. In its first form of organization, each worker carried out all the eighteen or so tasks necessary for producing a pin, starting with a coil of wire and ending with the polishing and packing of the finished pins. Smith then discussed a second form in which the workforce was divided into groups, each of which would be responsible for one task in the process of pin-making. His observation was that this division and specialization could only result in greatly increased productivity.[2]

Specialization and division of labour are two principles of organization which have characterized human development. The influence of these principles on production has been immense because it enabled differences in aptitude and skill to be used to the benefit of the community and greatly increased its economic activity. Their influence also underlines the extension of social interaction within a community: a community in which specialization has not emerged can be thought to interact at a level expressive of basic concerns, such as the need for shelter, security, food and procreation. Specialization adds to this range of feelings the notion of discussion of the functions of an individual or group of individuals.[3] This concern with the value, or instrumentality, of skills points also to the concepts of control and coordination because these provide at least one means of seeking to maximize the contribution made by such specialists to the social and economic health of the community.

The work of libraries, information services and resource centres is the management of the interaction between staff, stock, users and technology. One of the tasks of management is to create an organization within which this interaction can be achieved efficiently, effectively and economically. There are several factors which can be adjusted, or controlled, in order to affect the quality of this interaction. The balance between provision of staff and the provision of technology is one such factor; the balance between expenditure on stock and expenditure on personnel is another. The form of organization may also vary from highly fragmented (many departments, many lines of communication, many lines of control, sharp distinction between policy-making and administration), to undifferentiated (tasks shared, group decision making, group responsibility). Such differences are to some extent a reflection of personnel numbers, range of services and distribution of service points. They also reflect management styles and the organizational culture of any parent body of which the library, information service or resource centre happens to form part. This diversity again underlines the need to understand the unique characteristics of an organization before suggesting changes or proposing solutions to problems.

Analysis of needs

For managers of libraries, information services and resource centres the most familiar type of needs analysis is associated with the assessment of users' needs for information in order to design or improve systems for the delivery of information. In practice it has proved difficult to uncover the reality of needs for information because such needs are bound up with aspects of the psychology of individuals: what can be done is to infer needs from the actions of users and an understanding of their expressed demands.[4] The same can be said to be true of the study of information needs of groups such as staff, non-users, parent organizations and others with some interest in, or link with, libraries, information services and resource centres.

The method of insight and inference is also the only possibile way to study aspects of need other than that for information: the study of motivation, assessment of the need for training or staff development are examples where indirect methods of analysis and inference are often the only feasible methods.

Need is also associated with attempts to resolve disputes or rectify misunderstandings. This activity is an important part of the job of managers because such unresolved problems can have long-lasting and pervasive effects on an organization. Before taking decisions in relation to such problems it is imperative that the manager gain as full an understanding of the circumstances as possible. This includes trying to assess what the needs of the individuals and groups concerned may be. Once again, however, these needs can only be inferred. In addition, each of the individuals or groups may have a different perspective on the dispute or misunderstanding and an acknowledgement and understanding of each perspective is also required.

The traditional methods which have been used to study needs are largely based on the techniques used in the social sciences. Surveys using questionnaires, interviews or observation have been widely used and have provided some quantitative data describing categories of need; analysis of data derived from other sources such as statistics of use has also been employed. These techniques tended to produce a lot of descriptive data but often failed to provide any deep insights into need because of a lack of underlying theory. Because need is a very individual characteristic, seated in the context of the life of a person and reflective of many social, environmental and psychological aspects, it is largely obscured by the process of aggregation which is implicit in the quantitative approach.

A qualitative approach to analysis of needs has been developed which explores needs at an individual level, treating them as phenomena to be described and understood. The situations which an individual encounters, whether at work, in the home or in other social settings, are studied through open ended interviews and other participative approaches in order better to understand behaviour. What is sought is an enhanced understanding, a more holistic view, of individual need. As a technique, soft systems analysis forms part of the qualitative approach. Some quantitative techniques may also be used to complement insights and to seek whatever generalization of results may be possible, but the core of soft systems work is with the individual and small groups of people.

For a manager, the analysis of need is followed by a choice of action. Such choice is often difficult because the analysis will have revealed, or confirmed the existence of, several areas of need which cannot simultaneously be satisfied; in other words, there is a competition for scarce resources. Deciding to satisfy the needs of one group rather than another is rarely easy, especially if one group is perceived as being, in some sense, weaker than another. For example, should a manager always seek to satisfy the needs of the numerically larger group, thus pursuing a utilitarian approach? Is it preferable always to choose the approach which will use the least resources? Is it best to follow the path which will satisfy the most politically powerful group? These are not easy questions to answer and they are further complicated because some, but not all, of the characteristics of a group are perceptions rather than facts: political strength is not, for example, as objectively assessable as numerical size. Even the assessment of something seemingly as objective as the costs of a project may depend to some degree on point of view, because there are different ways in which overhead costs could be allocated or absorbed and there may be different opinions about how other costs should be allocated over a period of time.

Decision theory, which is sometimes suggested as a suitable technique for helping a manager to make a choice, can offer some assistance when the choice involves a comparison of factual, 'hard', data which, it is agreed, fairly represents the competing sides. It may even be useful when dealing with some of the perceptions if a consistent and coherent means of expressing them in a form for comparison can be derived. Two problems quickly arise, however: in the first place the reduction of perceptions inevitably distorts them and may encourage the manager to overlook, or discount, factors which could later assume great importance; in the second, such reduction may seduce a manager into forgetting that it is the aspirations of groups of people which are under discussion. 'Hard' analytic approaches may be very successful when used for dealing with technological problems such as deciding how many terminals to provide at an issue desk, but they cannot be expected to be successful when dealing with problems which are rich in perceptions and feelings but not in objective data.

This contrast between 'soft' and 'hard' data is typical of many of the problems which managers of all kinds of organization encounter, and libraries, information services and resource centres offer no exceptions. Their basic aim is to deliver relevant information to users in an economical and timely manner; the use of information technology has multiplied the ways in which this can be achieved and has enabled them to offer new and more effective services. However, this has also confronted their managers with the problem of designing and managing the link between the 'people' side and the 'technological' side. Designing one in the hope that the other will, somehow, fit or adapt is not a successful strategy, but conventional approaches to planning and design sometimes encourage managers and designers to behave as if this were possible: there may be a suggestion, for example, that the best strategy is to dissect the planning or design problem into separate categories, explore each and then consider a combined solution. The focus of attention is often on the technological aspects because these seem to be more tangible and substantive than the human aspects.

A word of warning is also necessary here. What is being suggested is that approaches to planning, management and design issues which neglect the people aspects of an organization are unlikely to be successful. It is not suggested that all approaches which use 'hard' analysis or techniques are wrong or that the products of their use should be ignored; rather, it is proposed that such approaches can, and should, be used where appropriate and that they should be complemented with a type of analysis which highlights the human dimension whenever technology and people meet.

Systems and systems thinking

'System' is a word which is frequently encountered nowadays in all sorts of contexts. It may be used to describe tangible things such as parts of the body (the lymphatic system), physically remote things such as astronomical objects (the solar system), abstract concepts such as a legal system and technological entities such as a computer system. It is also used to describe some homely objects such as a new type of washing powder or a diary. 'System' is such a ubiquitous word that one could well ask if there is anything which cannot be described as a 'system'. Rather unhelpfully, the answer seems to be 'no', because a system consists of things, ideas and actions which are linked together in some way and, if one looks and thinks hard enough, it is always possible to perceive some kind of link between components. If this approach is pursued, 'system' is no longer a useful word because it is descriptive of everything and particular to nothing.

A pragmatic way of resolving this problem and rescuing the word is to confine its use to describing sets of components which are linked together for some purpose and about which it is sensible and helpful to talk as if they are a whole. So, for example, it would make sense to discuss the national interlending system, recognizing that we mean by that the libraries, processes, users and documents involved in the scheme but not components such as the shelves on which the documents are stored or the paper used in the fax machines for sending requests. These latter components are certainly present somewhere within the interlending system but it is difficult to imagine a sensible discussion which would mention them in the same sentence as one discussing libraries, processes, users and documents. They are more tangible, more particular, than the concepts of libraries, processes, users and documents. On the other hand there might be occasions when it would be sensible to mention storage components or fax paper: we might be discussing the shelving system of a library or a purchasing system for consumable items.

Some writers have adopted the word 'holon' as an alternative to 'system', arguing that the former should be used to identify and discuss 'abstract ideas of a whole having emergent properties',[5] properly the province of those involved with systems thinking, whilst the word 'system' should be confined to everyday use. This idea is certainly attractive and has been followed by some writers especially concerned with the epistemological basis for the study of systems; however, in this book the use of the word 'system' has been retained because the focus is practical and the use of an unfamiliar term could confuse rather than clarify.

The conclusion is that knowing when it is useful to employ the word 'system' is a little like recognizing groups in classification theory: the components need to be similar in their level of abstraction. In addition, because of the flexibility of use of the word it is always necessary to make sure that everyone using the term in a discussion has agreed on what comprises the 'system'. Quite different approaches might be taken to discussing the national interlending system, for example, if one group included users in the definition whilst another did not.

There is another similarity with classification theory: the notion of hierarchy. In a classification schedule, a 'thing' may be divided into its 'parts' or a process into individual actions. As one moves from one level of abstraction down to another, the level of detail becomes greater, almost as if one were looking at the 'thing' through a microscope at higher and higher stages of magnification. The higher the degree of magnification, the smaller the immediate field of view and the less perception the viewer has of the whole 'thing'. In a similar way, a system may be represented by subsystems, each of which can, itself, be considered as a system composed of subsystems. Just as with developing a classification schedule, or using a microscope, there has to be a practical decision about when to stop. Unfortunately, this is not something about which it is easy to provide rules: rather, it is a product of experience, of being able to recognize when a level of sufficient detail has been reached and any further subdivision will simply lead to confusion.

The ability to work at different levels of abstraction is important for any manager: the development of policy often requires the manager to work at a high level of generality. The director of an academic library, for example, might be involved in defining a new loans policy and wish to deal with users within the library system in three categories, namely faculty, undergraduate students and postgraduates. Subsequently, a problem with loans to postgraduate students in the medical faculty could alert the director to the need for special arrangements to be made for this particular subsystem of the academic library system. In developing the rules for that special arrangement the director would also have to take into account the policy which has been defined for the 'higher' system. Systems thinking encourages such flexibility of mind and also prompts those who use it to consider the effects of any decisions on the various levels which may be recognized in a system.

Another important characteristic of systems thinking is its encouragement of multiple, but connected, 'views'. Branch libraries might be considered, on the one hand, as a network of service points and, on the other, as a physical set of buildings requiring maintenance. Both views are valid and can be useful for different purposes but the connection between them is also important: a decision to close one building for major refurbishment may have a considerable effect on the quality of service offered at the other branches if users are diverted to them. Systems thinking encourages managers to consider the implications of the connections between such multiple views.

Two final points concerning holistic thinking. None of the 'views' is complete: each is a partial view taken from some particular perspective. Provided there are no errors of perception each is true and will be helpful in building up knowledge of the complete subject being studied. By working with a particular view a distinctive set of insights can be gained, but in the end they also need to be related to the whole. The second point is that in identifying and discussing a system the observer is constructing a meaning. If one calls a computer network a 'data processing system', or a group of staff a 'personnel system', that is not to suggest that these 'systems' are really tangible. The components of the network and the people themselves are, of course, real but the rules by which they operate and the properties which, together, they exhibit are a product of thinking. The mental construct, 'system', is an abstraction connoting components, ideas, processes and events which can usefully be thought of as a whole. On another occasion it may be more helpful to take the same components, ideas, processes and events and see them as being part of another system: again, this will be a mental construct but one which takes a different perspective from that used to view the previous 'system'.

Types of system

When thinking about how something works it is often helpful to look at it as a whole, noting what it appears to do, and then to break it down into parts and examine each part independently of the others. In doing this, one recognizes that the function of each part depends on other parts and, in turn, also affects some parts. Although considerable knowledge may be gained from understanding how one part works in association with others, the knowledge of what the thing does will not be complete until the complete thing is, once again, considered with this new insight. We may recognize components to store electric power, for example, and these, in association with additional components to regulate the release of electric current, could constitute a power supply system. It is only when we put this knowledge together with knowledge about other components such as a bulb and a reflector and lens (an illumination system) that the idea of a battery-powered torch can be perceived.

Four major types of system are often identified:

- *natural systems* which consist of things which occur by act of nature, and which can sensibly be treated as a whole, having describable properties: these include human beings and animals, earthquakes, plants and galaxies
- *designed systems* which consist of things created by people, which can be concrete, such as machines, or include abstract ideas, such as index languages
- *social systems* which consist of collections of people interacting in some way, such as work groups, families, social groups
- *human activity systems* which consist of humans engaging in some purposeful activity, such as reading, discussion and sport.

Not all types will necessarily be present or recognizable in a useful form in all circumstances. For example, in describing the activities of a bridge club there are designed systems in the form of the tables on which the game is played, chairs for the players and the rules by which they play. There are also social systems in the relations between the players and between the committee who run the club and its members. The human activity system of 'a club at which the card game of bridge is played for entertainment' is the description which usefully draws these other systems together, but one could also identify aspects of natural systems such as considering the members as biological phenomena (flesh, muscles, blood, organs and skeleton) and the tables as processed botanical phenomena (organic materials). This extension would not, however, be very useful unless one were concerned about the effect that playing bridge might have on the biochemistry of the members or with the consumption of rare tropical hardwood to make the tables.

One can also identify different systems in an organization. To take the example of a college library, we might recognize the personnel (a social system comprising the individuals who work there and the work groups to which they belong) who are managed through a personnel system (a human activity system keeping records of what personnel do, salary, and so on) and who manifest their activities through a human resources plan (a designed system consisting of rules for staff allocation to service points). There will also be other systems associated with technology – a computer network, for example – many other human activity systems, such as a user assistance service and reference service, which also make use of other designed systems such as catalogues and classification schemes. Taken together, these form part of a larger human activity system, the college library, which itself forms part of a still larger human activity system, the college itself.

Systems are constructs of our minds, concepts which are useful when we are thinking about the relationships between human beings working together and between them and the technology or other designed systems which they use. Soft systems analysis is especially suited to assisting us to understand what is going on in a human activity system.

Systems analysis

The RAND Corporation is credited with having fostered the development of systems analysis, which is often seen as being a development of an earlier technique called 'operations research', sometimes abbreviated to 'OR'. The impetus for the development of operations research was the need to make the best use of resources in manufacturing, this being a critical factor during and immediately after the Second World War. By applying the methods of scientific analysis to the operations involved in manufacturing it was possible to develop an understanding of the process and to make mathematical comparisons of various alternative methods. Quite often such analysis could not be complete because what was being observed was an actual process of manufacturing, not an experiment under controlled conditions. Techniques of approximation had to be developed which allowed operations researchers to make the best use of available data to refine a process without necessarily being certain, or able to claim, that they had solved all of the problems.

This notion of approximation, of trying for the 'best fit', is characteristic of the idea of exercising a craft skill and is an important concept underlying all such intervention in real, rather than experimental, conditions. It is an approach which characterized operations research and which has also been carried over into systems analysis. Approximation does not mean, however, that a sloppy approach is either inevitable or acceptable: it is simply a recognition of the practical limitations on what can be done given prevailing knowledge, resources and time. It is a pragmatic way of avoiding the trap of 'analysis paralysis', where the responsibility of taking a decision is repeatedly evaded in the hope that further analysis and better data will reveal a unique solution to a problem.

Systems analysis developed an identity as a separate technique at the end of the Second World War,[6] with an emphasis on approaches to using technology to solve problems in industry and, later, within society. Successful use depended on there being a clear statement of objectives to guide the analyst towards ways in which those objectives could be met, and on the existence of a set of criteria that enabled a rational choice to be made between alternative means of reaching them.

This approach was quickly recognized as being successful when applied to problems which could be neatly summarized in a clear set of objectives and where a suitable technology was available. The application of computers to solving problems of data processing, such as the development of an automated financial control procedure, was a typical use of systems analysis. The objectives for such a procedure could be defined as 'deliverables', such as speed of processing, types of output and records to be stored. Criteria for choosing amongst the alternatives tended to focus on performance, economy and efficiency: the effect that the redesigned procedure might have on its users, the human component, was often perceived in terms of the need to train them. Questions about how the working lives of users might be changed by its introduction, how job satisfaction might be affected, often went unanswered or even unasked. The focus was on using new technology to accomplish a defined task and not necessarily on using it to improve working life.

Early attempts at using computers in the context of libraries, information services and resource centres tended to follow this same pattern. The task was to design a stock control system for such a context and analogies were made with the control of inventory in factories or supermarkets. The resulting systems displayed the limitations of the available technology as well as the restricted view which had been taken. For example, catalogues were produced in loose-leaf bindings, with each entry restricted to one line of 80 smudgily-printed characters, though little research had been carried out to discover if this form was either preferred or even acceptable as a substitute for the card catalogue. The focus was on redesigning procedures and redefining products so that the available technology could be used to effect economies. The deleterious effects that the limitations of that technology might have on users was largely ignored.

The techniques of systems analysis were also tried on problems which were social rather than technological. It was quickly recognized that technical problems were often susceptible to solution through the use of systems analysis because it seemed that they could be represented by objectives which were clear and unambiguous. It was also recognized that social problems could rarely be represented in such a coherent way. For example, finding the arrangement of shelving in an off-site library stack which maximizes the available shelf space is a technical problem which can be mathematically analysed; trying to design a shelving arrangement which maximizes shelving space and is comfortable and pleasing for public use raises such problems as the optimum size of gangways, choice of shelving heights in order to avoid creating a maze and placing some materials out of reach, logical order for the collection, need to provide seating areas and so on. Some needs may be conflicting: for example, the need to provide easy access to shelving and the need to provide ample reading space.

As it became apparent that the techniques of systems analysis could not be applied to social problems with the same expectation of success, a new type of systems analysis began to emerge. This came to be called 'soft systems analysis', the more technological form sometimes being referred to as 'hard systems analysis'. The essential difference between the 'soft' and 'hard' varieties is that the latter can be used successfully when dealing with problems which have been clearly defined, which lack ambiguities and where the objectives for the outcome are agreed; soft systems analysis is used to investigate problems which lack such clarity or where the nature of the problem is not well understood. This distinction also encapsulates another point: 'soft' is not necessarily better than 'hard', nor has it superseded it. Rather, 'soft' systems analysis may be seen as a method which can be used to investigate a problem which is unclear and, following this analysis, certain elements of the clarified problem may be seen as suitable for further analysis using 'hard' systems analysis. David Patching summarizes this difference in the methods: 'Hard systems thinking and analysis is essentially concerned with the question of *how* to achieve a predetermined aim; soft is concerned with defining the options for improvement, in other words, addressing the *what to do* question.'[7] Soft systems analysis is also firmly rooted in the examination of problems which concern, or have a large component concerned with, human behaviour.

Methods of analysis and the 'deliverables'

In discussing the analysis of needs it was noted that the methods of investigation tended to focus on the needs of large groups and rarely allowed a glimpse of individual needs or the needs of small groups. In other words, the 'deliverable' produced by this type of analysis is a statement of need inferred from pooling the views of many people.

The deliverables from hard systems analysis are answers to procedural and technological problems which have been clearly defined. In particular, hard systems analysis focuses on economic consequences and the efficient use of resources.

From soft systems analysis, the deliverables are answers to questions concerned with behaviour, attitudes, feelings and needs at an individual or small group level.

Each method has a particular role to play and can provide complementary aid in developing a fuller understanding of a problem and seeking a solution. A few problems may be encountered which need only one method to provide satisfactory understanding but, with the increasing use of technology and growing awareness of the responsibility of management towards the needs of individual members of staff, the likelihood is much greater that in most circumstances a combination of methods will have to be applied.

What characterizes success in the management of organizations today is the ability of the organization to be responsive to the changing needs of customers, clients and workers whilst at the same time continuing to operate economically. For an organization to be effective it must satisfy needs but it must also use its resources in the best way possible to meet those needs. Dealing with change is, therefore, a constant feature of the job of a manager and one which requires an ability to understand the work of the organization at many levels and to appreciate the role that the use of technology may play in achieving its objectives.

For managers of libraries, information services and resource centres, dealing with change is also a constant feature of life – but with additional complications. The objectives of libraries, information services and resource centres are difficult to encapsulate in a few brief phrases: they serve users with often widely divergent sets of needs and with no clear way of estimating their success in meeting those needs. The question of *what* to do is, then, often unclear and may evoke different responses from those responsible for deciding. Because the soft systems approach to problems recognizes and encourages the study of such differences of opinion and perceptions it offers a good starting point for developing strategies for the management of change in libraries, information services and resource centres. The essence of the approach is to encourage those who are concerned with these needs to speak for themselves, to share their own interpretations and perceptions of events, rather than allowing others to impose a meaning on them.

The management of libraries, information services and resource centres also throws up another complicating feature: the nature of the relationship between an information worker and the user of information. The information worker is, at one and the same time, an employee of an organization providing resources for the running of the service and an intermediary making the resources available to the user. Both are professional responsibilities but they may, on occasion, conflict. Consider the information service which has a small budget for online searching of external databases not available on CD-ROM: consider, also, the client who describes a pressing need for very recent information which could be met by a search on such an external database. The information worker has to balance responsibility to the employing organization against the need of the client. The soft systems approach can be used to explore, as case studies, the issues surrounding such problems so that a common and agreed policy can be established in advance.

If the development of libraries, information services and resource centres moves more towards providing access to information than the use of locally-held stock ('access versus ownership') then the need to make decisions around such issues can only become greater. The sense of being 'caught in the middle' between client and organization will become a strong feature of life for an information worker.

References

1   Miser, H. J. and Quade, E. S. (eds.), *Handbook of systems analysis: overview of uses, procedures, applications and practice*, Chichester, Wiley, 1985, 29.

2   Smith, A., *An inquiry into the nature and causes of the wealth of nations*, Cannan, E. (ed.), New York, Modern Library, 1937, 4–5.

3   Katz, D. and Kahn, R. L., *The social psychology of organizations*, 2nd edn, New York, Wiley, 1978, 258.

4   Rohde, F. N., 'Information needs', *Advances in librarianship*, **14**, 1986, 50.

5   Checkland, P. B. and Scholes, J., *Soft systems methodology in action*, Chichester, Wiley, 1990, 22.

6   Checkland, P. B., *Systems thinking, systems practice*, Chichester, Wiley, 1981, 134.

7   Patching, D., *Practical soft systems analysis*, London, Pitman, 1990, 27.

Chapter 2

The management of planning

PLANNING IS A major part of the work of most managers; for some it is the process which occupies the greatest proportion of their work time. It is also the process which is most highly esteemed and around which a particular mystique has gathered. The number of texts and journal articles dealing with aspects of planning attests to the important place it holds.

Planning occupies this position of prominence because of the increasing understanding individuals have of their environment. With understanding comes the sense of control and a recognition of the benefits of anticipation. If we were living in a world where there was little predictability, where we could not depend on knowledge of what happened yesterday, and has happened again today, to give us an idea what is likely to happen tomorrow, there would be little point in devoting any of our energy to planning. It is only reliable knowledge of past events and their consequences, the ability to spot patterns of events and a fair degree of certainty that those events will recur with similar consequences, which makes it possible to plan. Even so, it is only worth devoting energy to planning if this will render some advantage or enable us to avoid a penalty.

Knowledge of patterns and an assumption of predictability are, therefore, key factors in planning. However, another important principle also needs to be recognized: each person's perception of patterns of events is different and is influenced by a range of factors including personal circumstances and personal history, surroundings, experience and knowledge of previous events. It is no surprise to learn that the witnesses at the scene of some event recall it in different ways, sometimes contradicting each other and each person recalling different aspects. In addition, the events which surround us are so rich in stimuli and content that we all abstract particular things, cir-

cumstances and aspects from our perceptions to create models which
are simpler than the reality.

What goes on in planning?

Consider the task of planning the physical layout of a new department. The overall space allocated to the department may already have
been decided and its location and shape already determined by the
layout of the rest of the building. In other words, the plan must fit
into a predetermined pattern but within that pattern the space can be
used as seems best. The number of possible arrangements within the
space could be very large and these have to be reduced in some way
in order to make the task of selection manageable. An obvious and
almost intuitive step is to consider the way in which the space will be
used and from this to derive a criterion which can be used to limit
the possibilities. Perhaps the space is to be used for some task where
natural light is important: in which case, only those arrangements
which will position work surfaces close to windows need be considered. Even so, the resultant range of possibilities may still be large
and other criteria may have to be applied until a small set of possibilities remains.

What has gone on in this example is a process of model-building:
the actuality of the space is reduced to a two-dimensional outline and
particular factors, such as the source of natural light, are then used as
major features of the model. It is easy to compare arrangements and
try different layouts until one is selected which fulfils the exact criteria established for the decision. Previous experience is used to predict
the success of each arrangement in meeting the criterion of maximiz-
ing the availability of natural light. The level of abstraction of the
model may not, however, always be suitable: if it were to be recognized that the angle or intensity of natural light falling on the work
surfaces was also a vital component of the choice then the
two-dimensional model might have to be expanded into three
dimensions in order to explore this aspect.

Notice also that the model, because it is an abstraction or simplification of reality, excludes all features except those which are purposefully included. Models are defined by the characteristics of reality which have been chosen to be represented: this means a model may sometimes lack information which could prove critical to the decision but the absence, and importance, of such features must be recognized by the planners. In the example above, the positions of opening, as opposed to fixed, windows in the original space might have a significant effect on the choice of layouts but would not be included in the model unless the potential importance of this had been recognized and the need anticipated.

Even the most careful planning may not anticipate the need to examine all characteristics, however. Perhaps the users of the work surfaces will be highly pleased by the ample natural light but rather distressed by the rainwater which leaks from the opening windows. A previous user of the space could have told the planners of this problem but this person was never consulted and could not, therefore, add an additional point to the model.

A central problem with the process of planning is clear: it depends on an interpretation of past events and that interpretation can only be an incomplete and partially correct view of reality. This is not an argument for rejecting the notion of planning, but for recognizing the inevitable limitations of the process and for trying to ameliorate the worst effects of those limitations.

The planning cycle

The process of planning is sometimes represented as a cycle of activities, often called the 'Planning Cycle' (Figure 2.1). As a starting point for discussing the process of planning the cycle is adequate: when reviewing a simple process of change in an organization each of the activities can usually be identified and the sequence followed.

By 'simple process of change' I mean a change which affects only a small part of the organization and only a few people. For example, the purchase and installation of a new photocopier to service the

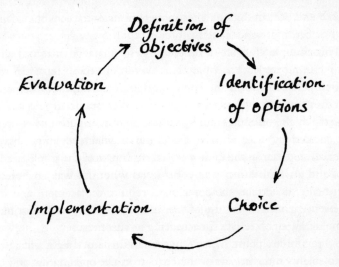

Fig. 2.1 *The planning cycle*

needs of the staff of an administration department would involve one person or a small group defining the performance required (should it be capable of colour reproduction, enlargement and reduction of images, automatic double-sided copying?). The result would be a requirements specification. Consideration would also have to be given to what criteria should be used to make a choice because there could be several photocopiers with a range of facilities capable of meeting the specification; thus, price per copy, cost of maintenance and supplier reliability might be selected as suitable criteria. Local suppliers could then be contacted to obtain details of suitable photocopiers and a final choice made after comparing the specifications to discover which copier most closely meets the criteria. If there were several which met the criteria in similar ways the final choice might depend on a comparison of other features of the contending copiers, such as intervals between maintenance, ease of dealing with minor running problems such as paper jams and so on. The final stages would be to conclude an agreement with the chosen supplier, make arrangements for installation such as final choice of the site and the provision of power supplies, and then to check that the equipment offered and the conduct of the supplier meets or exceeds the advertised specifications and agreement.

This example illustrates a simple process of change in accord with the planning cycle. The objectives were defined in terms of the desired performance of the copier and recorded in a requirements specification; the options were identified by gathering details of copiers which could meet the specification. Criteria for choice were also decided and used to make a decision on which copier to obtain. Practical steps for installation were taken as part of the implementation and an evaluation was conducted when all was complete. Eventually, no doubt, the copier will require replacement and the experience gained in the use of the existing machine will contribute to the setting of objectives for acquiring its successor.

To reinforce a point: my statement that this is a 'simple' change is based solely on the impact of the change on the organization and on the number of people involved. It is not necessarily connected with the technology involved: the comparison of specifications of copiers might well have meant coping with quite complex figures and statements but, provided the people doing this have the requisite expertise or access to competent advice, the process of change in this example is not intrinsically complex. The impact of the change is closely defined by the objectives which have been set: the provision of facilities to give a defined service of a particular level of performance within a particular area and to a specified group of people.

Planning and the mission statement

The process of planning has one purpose: to enable an organization to choose the best direction or action to meet a particular objective or set of objectives. In making a judgement that something is the 'best direction' or 'best action', it is necessary to assess it against a set of criteria which have been agreed to be appropriate. The criteria reflect such aspects of the work of the organization as its use of resources and its impact, and must express them in terms which allow comparison of possible directions and actions. In choosing a photocopier, for example, the objectives might be the ability to offer a low-cost and fast service; appropriate criteria might be cost per copy and speed of copying, with the objective being satisfied by choosing the copier with the best combination of high copying speed and low cost per copy.

By carrying out the process of planning, four 'deliverables' should emerge: a tactic, a strategy, a plan for implementation and a policy. A policy is the end point of the process of planning: it will incorporate a strategy, which is a set of tactics, and a plan for implementation. The mission, or aims, of the organization should provide the essential framework for policy by identifying its purpose, the groups and individuals it seeks to serve and other stakeholders for whom it has some responsibility or who have an interest in it.

Many texts on the management of libraries, information services and resource centres lay emphasis on the importance of a written mission statement which clearly identifies the intended role and the user groups to be served. The task of creating such a statement is often seen as being an important duty for senior management because it summarizes their agreement on the purpose of the organization and also reflects their views of the existence and needs of the various user groups identified in the statement. In essence, the mission statement forms one of the main contributions to the 'world view' taken of the work of the organization by the senior management team.

As we shall see in Chapter 4, the processes of revealing and exploring the 'world view' are an important part of the soft systems method; we shall also find that there is rarely one world view and that each identifiable group of stakeholders may hold views which are quite different. For the moment it is simply necessary to note that the mission statement represents part of the world view taken by the senior management team.

Whilst the aims of many organizations may be neatly and unequivocally stated in a mission statement, the same is not always true for libraries, information services and resource centres. One of the questions used to promote discussion when creating a mission statement for a commercial organization is 'What business are we in?'; the aim of the question is to encourage the participants to think beyond the confines of the range of products being produced and towards what essential need they are satisfying. This wider view may result in a mission statement which encourages the company to expand its product range in order to satisfy the identified need in other ways. For many libraries, information services and resource centres, however, this fundamental question is hard to answer because the range of user groups which could benefit from the services and available stock is very large and may lack clear definition. Need for information is, as discussed in Chapter 1, not easy to analyse. The consequence is that the development of a mission statement may be tacitly avoided or, if it is undertaken, the result may be bland and unspecific.

There is another danger attendant on this task. The development of a mission statement depends on a sharing of information, ideas and opinions amongst the senior management team, followed by discussion aimed at clarifying and reaching consensus. Whilst blandness may be the result of a lack of depth in the ideas expressed, it is also possible that it results from a tacit wish to avoid conflict between members of the team. Conflict avoidance is typical of satisficing approaches where decisions are accepted which minimize disagreement: in this case, the wording of the mission statement becomes an exercise in satisficing and appeasement rather than a sustained attempt at examining and seeking to understand the differences in perception.

In my experience it is much easier to develop an effective mission statement if the environment of a library, information service or resource centre allows the clear identification of a few user groups, each having an equally clear purpose for using such facilities. If the range of possible users is very large or amorphous it becomes difficult to identify a specific set of services which will be suitable. For the information service of a company, or the resource centre in a school, to focus on the needs of particular groups is a much easier and more automatic step than one would find in a public library, for example.

Why should there be this difference? Why is it more difficult to express what a public library should be doing than the information service of a manufacturing company? Why should a large range of possible users make it difficult to define a mission? Part of the problem may arise from the nature of a mission statement, the essence of which is a crisp and concise account of the purpose of an organization. An information service is addressing the information needs of a group or small number of groups of people which arise directly from the work of the parent organization. The public library is attempting to address the information needs (and this term includes recreational reading) of an often large and diverse population which arise from their interests, ambitions, desire for entertainment or wish for education. This is a much richer and more diverse amalgam of needs and one, moreover, which defies attempts at succinct expression. It is also one which may engender argument because the resources available may be inadequate to address all of the needs: by formulating mission statements which dissect and reveal the full range of information requirements it may also become apparent that the needs of some groups tend to be supported to a much greater extent than others. Moreover, the formulation of an explicit mission statement may force senior management to acknowledge such tacit bias. Both results can be deeply unpopular and, in order to prevent such searching self-scrutiny, the senior management team may resort to accepting a more anodyne formulation.

Whilst, in itself, the satisficing approach is not necessarily bad – it is sometimes more beneficial to preserve the peace for the time being than to launch into a searching critique – it will have a deleterious impact on the process of planning as outlined in Figure 2.1. This is because the initial steps in the cycle – definition of objectives and identification of options – depend on the planners having a shared and clear view of the context for which the plan is being developed. The temptation is to focus on the events which have led to the need to plan some response, because these seem to be tangible and require an immediate answer, whereas there may be deep, unrecognized, and largely unexplored differences of opinion about the context. Similarly, the selection of appropriate criteria for making a choice amongst competing actions can also be confused by an inadequate understanding of circumstances.

To emphasize a point, I am not implying that the process of planning is necessarily inadequate or that the representation of the planning cycle is wrong; rather I am suggesting that both are based on an assumption of the existence of a clear, unambiguous and shared sense of direction for the organization. The process of planning, as represented in the planning cycle, will function well where the mission is clear, the objectives can be recognized and stated and the criteria for success are uncontested. Unfortunately, the difficulties in developing a mission statement for libraries, information services and resource centres can make this assumption problematic.

Hard systems analysis and planning

In Chapter 1 systems analysis was introduced and a distinction between 'hard' and 'soft' was briefly explored. In order to appreciate the role of both types of method of analysis it is helpful to consider the link between hard systems analysis and planning.

Early uses of what we would now recognize as hard systems analysis emphasized the power of applying mathematical approaches to the study of systems in order to improve efficiency. As more experience was gained in its use it was also recognized that efficiency, on its own, was not the sole criterion for success; the needs of the stakeholders, especially the users, involved in the system being studied also needed to be taken into account if an effective system was to emerge. At the same time, the range of problems to which the technique was applied broadened to include the analysis of social and political problems as well as technological. In recognition of this wider application the term 'policy analysis' was sometimes used to describe the process when the problem under consideration was primarily concerned with economic, political or social aspects of life. The failure to acknowledge that many problems in social and political contexts cannot be reduced to making the most economically efficient choice from a range of technical solutions is usually given as the reason why early applications of hard systems analysis to problems with a strong social dimension did not meet with unqualified success.[1]

The essential role of systems analysis is to assist those confronted by a problem or an opportunity for change to choose the best way to solve the problem or make use of the opportunity. This implies that there are options to be considered and that the choice of a suitable direction is not immediately clear. Systems analysis is, therefore, an important adjunct to planning in that its use should encourage a full exploration of the choices available to the decision makers engaged in a project.

It is often necessary to underline this role of encouraging exploration of options. One of the problems which may afflict managers is a failure to recognize that there are several routes to the solution of a problem or for making the best use of an opportunity. The failure to recognize options comes not from a lack of competence but an inability to step outside the framework of the immediate circumstances. Erich Jantsch[2] has summarized this problem in his contrasting of 'normative' and 'exploratory' forecasting. Normative forecasting is based on the assumption that what *has* happened is a good indicator of what *will* happen provided the conditions remain stable. Decisions made on such an assumption will necessarily tend to reflect prevailing policies and actions will be preferred which are similar to those taken previously. Exploratory forecasting takes as its starting point the strategies and actions currently in force but then considers what other choices might be available, particularly if the constraints on the present circumstances are changed.[3]

It is often tempting for managers to confine their decision making to a normative approach. Dealing with circumstances which are familiar and apparently well understood means that the risk associated with making a choice appears to be less because the outcomes should be more predictable. This must be balanced, however, against the often unacknowledged risk that the conditions within and outside of the organization may have changed to such an extent that new approaches are needed or should be tried. Systems analysis can offer many insights for the normative approach and its use also encourages exploratory thinking.

There is nothing intrinsically wrong with normative approaches to making decisions because there are many occasions when continuance of a previous pattern of actions is both appropriate and effective. It is, however, important for any manager periodically to question whether a fresh direction is needed.

In my experience, most problem solving by senior management should involve a mix of the normative and the exploratory approaches. The development and refinement of policy and strategies almost always implies a response to, or anticipation of, a change in circumstances sufficiently large to make sole reliance on normative approaches unwise. Decisions which are more concerned with administration and routine matters can rely more heavily upon normative approaches because they are usually taken within a framework of existing policy and strategy. Similarly, decisions about operational matters tend to be normative, being seen as an evolution of a set of previous decisions rather than being considered from first principles each time.

Judgement is also an essential component in all types of problem solving. Even though there may be a long pattern of previous experience upon which to draw, it is still necessary for the person making the decision to be sure that the conditions of that previous experience are still substantially in operation and that, moreover, the pattern of previous decisions resulted in satisfactory outcomes.

The distinctive characteristic of hard systems analysis is the belief that for many problems it is possible to define a present state and an ideal solution; the 'analytic' approach is concerned with an orderly and systematic investigation of the ways of moving from the present to the ideal state. The assumptions of this approach are that both present and ideal states can be described and that all decision makers will agree with the criteria used in making a choice. Whilst these assumptions may be true for some classes of problem it is also evident that there are many problems where there is little agreement on what an ideal state might be or even on how to describe the present state.

By way of illustrating these difficulties, consider the problem, introduced earlier, of replacing a photocopier: by describing the copying performance and cost per copy of the present machine we have a good description of the present state. By defining in similar terms what we think is needed we will have a description of the ideal state. A hard systems analysis approach will enable us to scan the market and select the machine which will most nearly approach this ideal state for the copying service. The problem is clearly defined and there are clear criteria for choosing a satisfactory solution.

Now consider what happens if to this technological problem is added a social dimension: suppose there is concern about possible infringements of copyright by some users of the photocopier who may try to make multiple copies of articles. Some of the features previously thought desirable might actually facilitate or encourage this kind of abuse and the discussion will move from a technical comparison of features and performance to issues of control and supervision. There might be disagreement on how the ideal state should now be defined: perhaps a supervised copying service would be a better way of providing a quick response to users whilst also adding the element of control. There may be disagreement on the context of the problem: perhaps it is not the job of libraries, information services and resource centres to concern themselves with preventing copyright infringement. There may even be disagreement on what the present state actually is: perhaps the existing machine seems quite satisfactory for making single copies and no replacement would appear to be needed for this purpose.

The addition of the social issue immediately prompts discussion of aspects which cannot be reduced to comparisons of technology or comparisons of performance. Instead, judgement, opinion and perception become important factors to be considered and debated. In addition, the problem itself has become less clear-cut: there is still a strong technological component but the criteria for recognizing an optimal solution are no longer immediately clear. There is also uncertainty about the objective to be achieved in solving the problem. Whilst it would still be possible to use hard systems analysis for investigating the technological aspects of the problem, that process must wait until the objective has been decided.

Hard systems analysis has an important contribution to make by providing a framework within which managers can be assisted with decision making but its strengths are contained in the systematic exploration of clearly-defined problems and using coherent criteria for making a choice. At the core of hard systems analysis is a model somewhat like that of the planning cycle. It is variously represented by different writers such as Cleland and King.[4] The essence is:

- systematic examination and comparison of those options which are related to the accomplishment of desired objectives
- comparison of options on the basis of the costs and benefits associated with each
- explicit consideration of risk.

From this model it is clear that the prime elements which must be in place in the minds of the decision makers are the objectives, the costs and the benefits which would accrue from each choice. Without clarity and agreement on these aspects it is difficult to apply the techniques of hard systems analysis with any benefit.

Bounded rationality

Cleland and King[5] also discuss the concept of 'bounded rationality', an idea first developed by Herbert Simon. The suggestion is that people often cope with decision making by adopting a satisficing approach: they seek a solution which will be 'good enough' for their purposes and stop the search there. A more rigorous search might well suggest other solutions and ones which, in retrospect, would have been better. The approach of 'bounded rationality' is intuitively acceptable to many (perhaps most) people. To define and evaluate all possible courses of action will often take a very long time and will be expensive: the cost of complete rationality has to be set against the benefits which might accrue from such an exhaustive search

Introducing the 'bounded rationality' principle into the model of hard systems analysis presents a problem, however. The model operates on a 'stepwise' principle whereby each stage is completed before the next is undertaken and the results from each stage should be clear and uncontested. 'Bounded rationality' implies, however, a definite limit to the depth and range of analysis and the possibility of disagreement over when the search for tenable solutions should be concluded. In other words, who is to set the 'bounds'?

Even within the domain of technological problems such as the specification and introduction of new computer systems, there has been discontent with the traditional hard systems analysis approach, and extensions of the approach, such as Structured Systems Analysis and Design Methodology (SSADM), have been developed. The dissatisfaction with hard analysis as often practised can be attributed to the tendency to ignore problems of definition and understanding in the early stages which then emerge later only to hinder further progress. The result is often that failure to deal effectively with the initial problems leads to the development of solutions and products which do not meet the real needs of users. What is needed, instead, is a method of analysis which distinguishes between the context of a problem, the logical design of a solution and the physical design of the means of overcoming the problem.[6] The temptation, especially for creative people who are 'sure they know what to do', is to get on with the physical design as soon as possible and to neglect fully to consider what a solution must achieve; in other words, they may concentrate on the 'how' at the expense of the 'why'.

Systems analysis and social planning

The management of libraries, information services and resource centres represents a problem which is surprisingly complex. There are technological issues and developments to consider – the development and introduction of appropriate information technology is one example – and there are decisions to be made about groups and individual members of staff, users and those with unmet information needs. There are political and legal aspects also, such as adherence to copyright conventions and freedom of access to information. Social and ethical considerations also come into play, such as questions of censorship. Almost any problem or decision about an aspect of the management of libraries, information services and resource centres will involve a mixture of technological, social and other considerations.

As we have seen, a starting point for analysis is usually to ask 'what are the facts?' The asking of this question, whilst obvious, implies a large assumption: that of shared understanding and shared experience which can lead to a single, agreed, definition.[7] In many social situations the primary problem is often to discover what each participant believes the facts to be and to discern whether there is any consensus around which further discussion could coalesce.

There may be less of an apparent problem if the discussion involves just senior management from within the same organization. The shared culture of the organization may be strong enough to encourage a single view, or at least a set of views which are not fundamentally different. However, the service culture of libraries, information services and resource centres encourages the development of a user-centred approach and to be effective this must involve both managers and users in a dialogue. The notion of a shared organizational culture disappears and tacit agreement on such concepts as the educative role of public libraries cannot be expected.

Even within a single organization or a single managerial team the assumption of shared understanding is often unsafe. All groups develop norms of behaviour around such matters as displaying agreement, dealing with disagreement and deferring to the authority of a group leader. Some groups may develop norms which stifle disagreement or which encourage their resolution through private discussion rather than open debate. Although there may seem to be open discussion and an appearance of negotiation, what is happening is a concealment of disagreement in favour of an apparently unified approach.

Case study

A small case study illustrates the point made in the previous section. Some years ago I was asked to advise on the setting up of an information service for a large charity. The service was intended to supply the various sections of the charity with up-to-date information about the work of this and similar bodies, and to gather published and semi-published information about the specific social problems the charity sought to address. The invitation to advise had come from the director of the charity, a person who had taken on the post following a distinguished academic career. The starting point for my involvement was a briefing session with the director. Having outlined the objectives of the charity and its management structure, the director spoke about the purpose and role of the information service, emphasizing the importance of keeping everyone who worked for the charity informed of its various activities and progress. Although there was already a news sheet, produced by the division responsible for public relations, which was sent to those who gave to the charity on a regular basis, the director did not believe that this was a satisfactory vehicle for keeping the staff of the charity informed. Having seen a few specimen copies I tended to agree with this view but was still unsure why the director should believe that the expense of setting up an information service, however small, could be justified.

The next stage was to attend a meeting at which those in charge of the sections could explain more fully what they thought was needed. At a previous meeting the director had suggested that the charity should develop an information service and had asked all the participants to prepare a list of what information their sections required and wished to receive on a regular basis.

Some meetings are very lively and there is a freedom and frankness in the exchanges which keeps everyone interested and involved; disagreements surface but are dealt with through vigorous discussion and there is general good humour and a strong sense of the group moving together towards an agreement. This meeting was not of that type; it was formal, with each person presenting a short paper and with little discussion. At the end the director, who was chairing the meeting, summarized the papers and concluded that everyone was in favour of the idea. No one questioned this conclusion or suggested that it might need further consideration. I was invited to take the papers away, consider them and return with my advice on setting up a suitable service.

I had been uneasy during the meeting because, whatever the participants had said, they had not behaved as if they agreed with the idea. In order to test this perception I proposed to interview each section head separately, ostensibly to gain a more precise idea of the information needs of their sections. During the course of these individual discussions each volunteered that their section did not have much to gain from an information service and that their present resources were satisfactory. 'However,' said one of them, 'I'll go along with it because the others want it.' Somewhat frustrated by this time, I asked what her opinion would be if she learnt that there was little support for the scheme amongst the other section heads. She replied that the scheme would still go through because the director wanted the information service. 'You see, he's been an academic, he's used to having information around and being supplied with it whenever he has to prepare a speech or give a paper. Without a regular supply of information he feels he's not in control.'

Whatever one may feel about the attitudes towards the supply and use of information that this case study reveals, the crux of the matter is that a project to develop such a service would be bound to fail if it were conceived on the understanding that it was desired by all sections and would serve their interests. What agreement there seemed to be about the need for the project had been imposed by the leadership position and authority of the director.

Landscapes and views

As we have discovered, the sharing of information and knowledge is a necessary preliminary to developing a solution to most managerial and social problems.

To take an analogy, imagine standing with friends on the top of a high hill. Each person can move around and look in different directions and together the group can agree on what can be seen. They can pick out groups of buildings below and, perhaps with the aid of a map, agree on which villages are there. This can be done collectively, as a group, and once the correct orientation has been established there can be little disagreement on the fundamental shape of the landscape.

Now look again. Because each member of the group on top of the hill has some special interest in the natural surroundings, each is able to provide the others with a different 'view' of the landscape. The ornithologist can point out nesting sites for birds; the geologist can explain what forces have shaped the local terrain; the hydrologist can suggest how to improve the collection of surface water into reservoirs. Each 'view' has special characteristics and each serves to build up a better understanding of the landscape.

In the case study, it can well be said that there was no opportunity for the sharing of views; it was even uncertain whether there was any agreement on the landscape or, indeed, if the section heads and the director were standing on the same hill.

In choosing an approach to systems analysis, what is needed for many social and managerial contexts, including that of libraries, information services and resource centres, is an approach which prompts debate about objectives and directions, which encourages all participants to bring in their special knowledge and views, and which facilitates the expression of disagreement as a preliminary to its exploration. Above all, the approach must discourage acquiescence and the suppression of disagreement. It should deter managers from plunging into design and 'how to' discussions before the question of 'why' has been resolved.

Do all problems need systems analysis?

The short answer is 'no'. As suggested above, the benefit of systems analysis comes from a better understanding of the problem and this can only be achieved if there is sufficient time and commitment from those involved to carry out the analysis properly.

'Time-sensitive' decisions most clearly place themselves outside the analytic forum. The need to make a decision very quickly almost always precludes extensive information gathering and participative approaches. That this is rarely a feature of managerial life is not of course unique to libraries, information services and resource centres. Perhaps the circumstances which would best approximate to this context would be a sudden disaster, such as fire or flood. Quick action will be needed to save life, stock and buildings. Even here, though, a well prepared management team will have worked through a process of systems analysis in the form of developing a disaster plan: a successful plan will limit risk by reducing the need to take snap decisions. There are also many examples of the use of case studies and simulation in other areas of managerial life which are undertaken with the aim of improving the capacity of managers to respond effectively to the need to make such time-sensitive decisions. The aim is to experiment with different approaches and to consider consequences in a more relaxed atmosphere than would prevail if the emergency were real.

Many routine decisions do not require repeated analysis because they occur within a framework of previously agreed policy and procedures. There are also many trivial day-to-day decisions which do not need extensive analysis because the penalty associated with getting them wrong is insignificant or because the matter can easily be set right.

The areas which can benefit from the use of systems analysis are those where there are large risks associated with making the wrong decision, where there is uncertainty or lack of knowledge about directions to be taken and benefits to be gained, or where there are strong and competing views of what is needed.

An allied question is whether all problems need soft systems analysis; again, the answer has to be 'no'. The main value for the manager of using soft systems analysis lies in the assistance it can give in discerning and gaining a better understanding of a problem. A crude distinction which is sometimes drawn is that if a problem is already clear and the objectives well defined then one can move immediately to hard techniques; it is only if there is a lack of clarity or dispute about objectives that the soft approach is needed. This distinction suggests, too, that there is a definite sequence of application, with soft techniques usually preceding hard.

Practical observation of people and groups attempting to solve problems reveals, however, that these distinctions and sequences are rarely so evident. Although in studying or writing about them it is convenient to consider one and then the other, their use is often much more integrated. The reason for this arises from the behaviour of people trying to solve problems and the notion of bounded rationality. We have already noted that an exhaustive exploration of the full range of responses to a problem is unlikely because of the effort and costs involved, and that decision makers will often adopt a satisficing approach in order to limit the range to a workable extent. Having done this it is entirely possible that whilst using hard techniques a particular aspect of the problem is revealed as being unclear or contested. It is at this point that the management group can use a soft technique in order to gain better understanding. Having made some progress the group can revert to the hard technique and continue with the development of a solution.

There are also classes of problems which do not necessarily need hard approaches. Hard approaches have tended to be associated with the use of technology and particularly with the development and implementation of computer systems. For managers of libraries, information services and resource centres a large proportion of their work is concerned with people, whether staff, users or other stakeholders. The political and social components of this work are aspects which are especially suited to the use of soft systems analysis. Problems associated with people are rarely simple because of the range of responses to events, differing perceptions and the psychological complexities of understanding motivation and behaviour. Whilst soft systems analysis cannot pretend to provide a complete answer for the management of such complex matters in all circumstances, it can provide a useful tool for gaining a greater understanding.

The case study reconsidered

Would soft systems analysis have been of help in dealing with the case study presented earlier in this chapter? My initial reaction is to suggest that it would not have been of much assistance, because for any method of analysis to work there must be a commitment to the process of analysis and a recognition that what might come out of the process is not necessarily what one might expect or hope. In other words, being committed to a process of analysis means also being committed to the idea of change. The director of the charity was perceived as being interested only in the establishment of an information service and appeared to be unprepared for the eventuality of this service not being perceived as necessary by the section heads. In turn, the section heads were not willing to be seen to be going against the perceived wishes of the director. There was no commitment, in other words, to an open exploration. Soft systems analysis would not have worked in these circumstances.

This underlines another point: preparation of all those involved in the process of analysis is an important preliminary to any type of analysis. This topic of preparation, or 'socialization', for the process of analysis will be explored in Chapter 3, but it is worth noting that the case study also reveals one of the greatest impediments to analysis and participative problem solving: a lack of trust amongst the participants. This does not mean that either the director or the section heads actively disliked or had suspicions of the others; rather it means that each was unwilling and unprepared to reveal their true feelings, and the reasons for holding them, to the other members of the group.

My mistake in dealing with the problem outlined in the case study was in initially accepting the statements of the director about the need for an information service and assuming that the group could move directly to planning such a service. Instead, what needed to be clarified was what activities the director had to undertake on behalf of the charity and in what way these activities needed to be supported. A useful approach would have been to focus on the objectives of the charity and, in discussion with the director and section heads, disentangle the declared need of the director for information from a possible potential need for information amongst the section heads. By concentrating firstly on the objectives of the charity there is a good chance that there would have been a broad measure of open and real agreement on what they were, and this would have formed a sound base for moving towards areas of possible disagreement and making use of the soft systems analysis approach. Some trust would have been established within the group and there would have been greater confidence over discussing more problematic areas.

As with any management technique, soft systems analysis has to be used within a supportive culture: it is necessary to be sensitive to the mood of an organization or group and to be aware of the distinctive characteristics of the personality of each person involved. It is not a panacea – it is not, indeed, any kind of cure – but if used in the right way it can be a valuable tool for helping towards an understanding of what is going on and why.

References

1   Checkland, P. B., *Systems thinking, systems practice*, Chichester, Wiley, 1981, 141–2.
2   Jantsch, E., 'Forecasting and systems approach: a frame of reference', *Management science*, **19** (12), 1973, 1355–67.
3   For a fuller discussion of the planning framework proposed by Jantsch, see Underwood, P. G., *Managing change in libraries and information services: a systems approach*, London, Bingley, 1990, 10–15.
4   Cleland, D. I. and King, W. R., *Systems analysis and project management*, 3rd edn, Singapore, McGraw-Hill, 1983, 87.
5   *Ibid.*, 86.
6   Nicholls, D., *Introducing SSADM: the NCC guide*, Manchester, NCC, 1987, 3–9.
7   An interesting paper which fully explores the implications of this assumption is Thompson, M. and Warburton, M., 'Decision making under contradictory certainties: how to save the Himalayas when you can't find out what's wrong with them', *Journal of applied systems analysis*, **12**, 1985, 3–34.

Chapter 3

Working with a group

SOFT SYSTEMS ANALYSIS is a method by which individuals and groups can look at problems and opportunities with others in the belief that the participants know many of the components which will contribute to a plan. It provides a framework for a *process* of enquiry and synthesis and, to be used properly, whoever is conducting the process needs to know something about working with individuals and groups. To provide a full discussion of the social processes which occur in organizations and between members of work groups is beyond the scope of this book; the interested reader will find there are many discussions of organizational behaviour which can serve as starting points and some books are mentioned at the end of this chapter.[1-3]

The role of the analyst

It may sometimes seem as if the analyst is the key person in the whole process of soft systems analysis. After all, it would seem that the analyst is conducting the work of enquiry, summarizing and assisting in the interpretation of the ideas, data and statements which are put forward. The analyst will also be responsible for producing a report and feedback on the findings. A reasonable conclusion seems to be that this must make the analyst the person at the centre of what is going on.

Whilst this is a tempting model to adopt, it is also in direct contrast to what an analyst should be doing; indeed, its use would threaten the success of the whole process. The true role of the analyst should be to facilitate the work of the other participants in the enquiry. It is they who will be conducting the enquiries, surveying and weighing the evidence; they are the only people involved who can provide an interpretation of the data and draw conclusions. Although I have used the term 'analyst' to describe the role of the person responsible for the framework, it is only for the lack of a better, more descriptive, term; what the 'analyst' should be doing is acting as the 'manager of the process of analysis'.

Analysts and consultants

The analyst may encounter a group with a problem or opportunity to discuss in three ways: as an external consultant, as an internal consultant, or as a member of the existing group. Each type of relationship has advantages and disadvantages which will affect the way in which the analyst can work.

The external consultant is engaged from outside the organization and agrees to work with the group within the bounds set by a contract or agreement. Contact with the group is usually mediated through someone inside the organization. The advantage of this relationship is the distance between the consultant and the object of discussion: this distance allows the consultant to reflect on what is going on from a more objective stance and may encourage a more searching exploration of the problem or opportunity. On the other hand, the consultant is unlikely to know the objectives and internal environment of the organization in much detail and may be unaware of subtleties in the relationships between group members. Time will have to be spent in providing some of this information and what has not been conveyed will have to be gathered by the consultant in the course of the work.

An internal consultant is usually someone who has been selected from another part of the organization to work full-time or part-time on a project. Because of prior knowledge of the objectives of the organization it may be assumed that an internal consultant has an initial advantage over an external consultant. In many circumstances this is so, but it is wise not to assume that the view the internal consultant has of the organization is necessarily correct or consonant with the views held by the group members. There is some advantage to be gained from an explicit identification of objectives even though all members of the group may appear to know and agree with them. The disadvantage of internal consultancy is the susceptibility of the consultant to attempts at forming alliances. These can subvert the process of analysis by encouraging the avoidance of conflict and limiting the depth of analysis in order to preserve a relationship.

Perhaps the most unenviable and, at the same time, most promising position is that of the analyst who has been selected from within an existing group. By virtue of membership of the group the analyst is aware of the objectives of the organization and of the nature of the problem. By virtue of selection the analyst is someone whom the group trust and whose skill they recognize. But the chosen member is also tied by social and organizational links to all the other members of the group and may be inhibited from providing the kind of feedback and guidance that the group actually needs in order to deal with the problem or opportunity.

Whether to select an internal or external consultant, or a group member, to assist with a project is always a difficult choice, further complicated by factors such as availability of people and costs. In addition, some projects may encompass aspects which are confidential or especially sensitive and, whilst one would expect external consultants to adhere to a code of ethics which would prevent them from gossip, the need to enforce confidentiality may be a strong reason for keeping the task internal. External consultants are especially valuable in facilitating discussion where objectives are unclear or contested, rather than where the minutiae of internal arrangements in an organization are under review.

It is probable that the first experience that most readers will have of using consulting techniques will be as 'facilitators' to the work of a group of which they are already members. A natural path of development, if they are effective facilitators and enjoy the role, is towards short periods of internal consultancy where they facilitate the work of a group of which they have not been part. Thereafter, the role of external consultant is a possibility as an occasional contract or as a full-time occupation. In either case the reader must bear in mind that, even though the consultant is not formally a member of the client's organization, a kind of evanescent membership pertains throughout the period of the consultancy and some share in the outcome persists.

What does an analyst do?

An analyst is a member of a group, the members of which have been brought together because they are thought to have knowledge, insight or experience which will be relevant to solving a problem or planning to make the best use of an opportunity. The distinctive contribution of the analyst is the management of this process but this should not be taken, by either the analyst or the group, to mean that the analyst is somehow in charge, or leader, of the group.

On the other hand, it is also misleading to think of the analyst as being utterly self-effacing, setting the stage and then being content to allow the other participants to play their parts upon it. The analyst has an important role to play in association with the group but it is a role which demands careful preparation and some experience if it is to be conducted successfully.

The core of the work of the analyst is in assisting an individual or a group better to define and understand a problem or an opportunity and then to assist in the development and implementation of a planned response. The focus of the analyst must therefore be on the process of achieving that understanding and developing the plan, and not on the end product of the plan itself. It is sometimes quite difficult when working as an analyst to maintain this sense of reserve, especially if one also knows something about the characteristics of the problem which is being discussed. Nevertheless, it is very important to hold back from a direct engagement, at least in the early stages of the process. There are two reasons for this.

The first arises from the perceived role of the analyst at the beginning of the process. At that stage the analyst is bound to be the centre of attention because it is the analyst who has established the conditions for the process and will, at first, be getting the process under way. Such a position is a strong power base and one from which considerable influence on the work of the group can be effected. The other participants, especially if they do not know each other well or rarely work with each other, will look to the analyst for leadership. If the analyst steps beyond the point of assisting with the process and becomes actively involved in the design of a solution, a condition of dependency may develop: the group members will look to the analyst for approval and support for their ideas and statements. The sense of dependency can go so far that the group finally appeal to the analyst to solve their problem or develop their plan: 'You've heard all that has been said: what would you do, if you were us?' At this point the analyst has moved well beyond the point of assistance with under-standing the problem or opportunity: the analyst is now in danger of appearing to own it and being seduced into trying to solve it or devel-oping the plan.

The second reason arises from the relationships of group members. The essence of the process in which the group is involved is a sharing of information, perceptions and ideas. Each participant has something to contribute to a solution but the making of such a contribution should involve everyone in the group: it should not be a dialogue between the participant and the analyst but, rather, a means of sharing. If the analyst is the focus of attention and exchanges, each member of the group is likely to begin to feel isolated from the other participants and to try to establish a link with the analyst. This may limit the contri-butions of all group members because they may be cautious about upsetting this tenuous relationship. No sharing of ideas is likely to take place and, most importantly, responsibility for the eventual project will not be clearly seen as being that of the group as a whole. What may emerge is that the participant who expressed the strongest opinion in favour of the project or spoke the most will be seen by the other group members as the person to be made responsible. This is unlikely to be a satisfactory solution in all circumstances, especially if there are still strong feelings about the project left unvoiced.

To use the analogy developed in Chapter 2, the primary job of the analyst is to ensure that the 'landscape' is properly surveyed and the individual 'views' discussed; once this has been accomplished it is then acceptable for the analyst to supply any additional specialist knowledge which might be helpful to the work of the group.

In practice, it is much easier for an external consultant to concentrate on the process of the analysis. Because an external consultant has an initial lack of knowledge of the objectives of the organization and the characteristics of the problem or opportunity being discussed, this position is a natural one to assume. The other participants will not be surprised or alarmed to find the consultant concentrating on listening to the discussion rather than contributing views about the subject. This position also offers the consultant an ideal platform from which to observe the reactions of other group members and to prompt clarification or to note differences of opinion. At the same time, it is important not to cultivate an air of Olympian detachment or to be so self-effacing as almost to disappear. The consultant has a vital role to play and must appear interested in the work of the group and involved with the details of the problem or opportunity.

For the internal consultant or the group member taking the task of analyst, this reserved position is difficult to establish and maintain, especially at the beginning of the process. When a group of people come together for the first time, even though they may be acquainted through other contexts, there is a period of socialization during which a role for each member begins to be recognized by other members and alliances begin to form. Individually and collectively the members of the group are trying to reduce the level of uncertainty by establishing guidelines, or 'norms', of behaviour for the group. This is an essential part of the behaviour of a group, necessary for its growth and effective functioning. An internal consultant or the group member taking the task of analyst may find it difficult to resist being drawn into such early alliances, but it is important to prevent the formation of overt, or even covert, links of this kind.

The difficulty is even greater if the group regularly works together on other matters. Norms of behaviour and roles will already be firmly established and, because these tend to be strong, it is often hard for the group to accept that one member will have to behave in a way which is different from that which they have come to expect.

Groups and socialization

A pragmatic way of trying to address this issue of the role of the analyst and, at the same time, to assist the group in its process of socialization, is to reserve a little time at the beginning of the first meeting in order for the group to consider how to work. It may be quite difficult to persuade group members to do this, especially if their natural concern for dealing with the problem or opportunity outweighs their interest in working together well as a group. The best approach is for the analyst to explain briefly that the business of the first meeting is to reach agreement on the purpose and timetable for the project and then to invite the participants to outline their interests in the project and to indicate in what way they are linked to the other participants. None of this information is necessarily challenging or contentious, though it may, depending on whether the group members have worked with each other before, contain some unexpected or previously unknown information. By inviting comment and questions from other group members after each contribution a process of communication will gradually be built up. Most importantly, everyone will have had something to say and will have contributed something to the general understanding.

During this series of short presentations the analyst can be gathering some useful information about how the group works, or is likely to work. Because the success of soft systems analysis depends very much on communication it is important that good patterns of communication be established as soon as possible. From simple observation it is easy to conclude that some members of the group speak more frequently than others, but this view needs to be enriched by other considerations. The apparently-silent person *may* have nothing to say, or may say little but speak with great impact.

Communication in a group: the case of the 'silent member'

As an example of the importance of considering the effect of communication patterns, another small case study may be helpful. A committee had been formed to consider the need for a particular range of facilities in an academic library building. I was asked to attend meetings of the committee, as an external consultant, because I was already assisting with some aspects of planning and implementation. The meeting took place around a 'U' shaped table and, there being plenty of room for the members around the table, we found ourselves spread out. The last member to arrive selected a place at the end of one arm of the table, sat down and turned his chair so that he could see everyone. He said nothing as the meeting progressed, took no notes and seemed to spend a lot of time looking out of an adjacent window. I concluded that he was either not interested in the business of the committee or so unprepared that he could not make a sensible contribution. Other members discussed the issue and, when it was evident that an agreement was in the offing, the chair of the meeting addressed the silent member and invited him to speak.

This he did with devastating effect. He presented a coherent and cogent analysis of what had been said, indicated the strengths and weaknesses of the arguments and then succinctly stated his proposal, which was in direct contrast to that to which the committee had seemed previously to be inclined. When he had finished, the chair put the matter to a vote and the proposal of the 'silent' member was accepted by a large majority. I enquired afterwards about this behaviour and was told by other members that this was a fairly regular occurrence, especially if a matter of some importance to the academic life of the campus was being debated. The 'silent' member, who was a senior academic, favoured the use of taciturnity in order to maximize the effect of his eventual contribution: a simple, and in this context, outstandingly effective tactic.

Whilst this type of behaviour may be rare it is important for an analyst to recognize when it is going on and to consider how it will affect the working of the group. If the members of the group come to expect this tactic to occur, and for it to be successful, this may inhibit the voicing of their own contributions because there will seem little point in their being made. The group will seem to be swayed by the strong argument and will vote in favour almost as a ritual. Their true feelings will be cloaked and the information and knowledge they might each contribute will remain largely unvoiced. Disagreement will, almost certainly, emerge later in discussions outside the group meetings and may have a serious effect on the completion of the project.

How should the analyst cope with this? A useful strategy is to emphasize at the beginning of the process that the first meetings of a group working with soft systems analysis are not primarily decision making meetings; though they may involve making some decisions about procedure, timing of meetings and so on, they are intended to be exploratory. This limits the likelihood of 'winning' and 'losing' sides being created and reduces the effectiveness of the tactic: a 'silent' member will lose opportunities to take part in the discussion and, because the meeting is not leading to a decision affecting the outcome of the project, there will be little to be gained from holding back.

Silence is not always, of course, an indication of subtle manipulation. It may also result from boredom, being ill-prepared or having lost the thread of the argument. It is difficult to cope with the first problem, though the policy of starting the process of enquiry with a position statement from everyone may assist: most people respond to an opportunity to talk about some aspects of their work. If boredom does seem to continue to be a significant factor it may be helpful to ask everyone to deliver a short position paper as a regular first item of each meeting.

If boredom in one individual is perceived to be a continuing problem, a private discussion with that person is a good idea. In general, this is the best approach to follow whenever the behaviour of one individual is disrupting or not furthering the work of the group. Clearly such an encounter needs to be handled with care and tact. If possible the analyst should check discreetly with the supervisor of the individual concerned to discover if there is some special reason, such as illness or personal circumstances, which might account for this behaviour. Above all, the analyst must avoid any suggestion of accusation or blame but must seek instead to understand how that person perceives his or her own role in relation to the project and the group. The analyst is not there to pass judgement on behalf of the organization or other members of the group but to identify ambiguities and encourage their resolution. Many instances of apparently unhelpful behaviour arise from individuals not being sure of their role or what is expected of them rather than from a determined wish to obstruct.

Patterns of speaking

Once the formal aspects of a meeting have been completed, and discussion is flowing within a group, the analyst may observe distinct patterns of speaking. A pattern where a statement by member A is frequently followed by a statement from member B is an important clue to the establishment of clusters of support or opposition. In the early stages of soft systems analysis it is quite important that the group and individual members remain as open as possible to their collective and individual ideas. Taking sides at an early stage is not a healthy indicator and if the analyst perceives a pattern of support or opposition growing it is wise to intervene. A suitable way of coping with this is to try to break the flow of the exchanges by asking a question immediately member A has finished or by seeking to summarize what has been said. This gives the other members of the group a short break in which to consider what has been said and the cumulative effect of

statement followed by support
or
statement followed by opposition

is lessened. The intention is to encourage the group members to concentrate on the content of each statement and not on the rhetoric. This should lead to a lowering of tension and less reactive behaviour.
Interruption is another important factor to consider when review-

ing communication. Power relations, which are considered in the next section, may be the reason why one member often interrupts another; this is most commonly seen when a group includes members who are not perceived as being of equal status either in terms of the organization or in social terms. Whatever the cause, interruption can be very destructive of relationships and of the process of analysis. People who interrupt others assume that their own contributions are of more importance than what they *believe* the original speakers were going to say. Not only has the group lost the possible benefit of the contribution of the original speaker: it has also lost the thread of an orderly discussion. If interruptions are frequent, not only will the meeting become disorganized but the member being interrupted may give up trying to contribute and withdraw into silence. This will foster a sense of resentment and may also result in important information being suppressed.

Coping with frequent interruptions in the context of a soft systems analysis meeting is not easy. The role of the analyst, as we have seen, is to facilitate and not to direct or control the meeting. Interruption many not necessarily be anything more than a wish to clarify or add details thought to be important at that point. The analyst must judge, however, whether the interruptions are having the effect of dissuading someone from speaking. If this is the case then the effects must be countered. To deal with isolated examples of interruption it is best to wait until the interruptor has finished and then remind the group that the person who was interrupted still has something to say. For more frequent occurrences it is best to raise the matter with the group deliberately and ask them whether they consider such behaviour to be a problem and how they wish to deal with it. It is advisable to approach interruption as a general problem of group behaviour rather than singling out an individual, unless repeated efforts to mitigate the effects of such behaviour have had no effect. The danger with focusing disapproval on the frequent interruptor is of causing a heated exchange and possible withdrawal of that person from the meeting. It is rarely a risk worth taking because it can cause resentment; sanction by the group is, in any case, usually more powerful and effective as a means of controlling individual behaviour.

Adopting formal rules of procedure, such as speaking only when invited by the chair, is a mechanical way of controlling behaviour which is unlikely to be successful in a meeting aiming to use soft systems analysis. Such meetings depend for their success on a ready exchange of ideas and views: they need an atmosphere of acceptance and closeness which is not consonant with a high degree of formality.

Power in a group

Another consideration arising out of the case of the silent member is that of the power relations between members of the group. A convenient way of looking at this aspect is to consider the 'power bases' which may form the foundation for whatever influence on a group a particular member is able to exert.

Five bases of power are conventionally recognized:

- **Legitimate power** comes from a position of authority in relation to other group members; since it is inherent in the organization and not the person it can also quickly change. Legitimate power is often associated with the power to reward or punish.
- **Reward power** comes from an ability to bestow a benefit on a group member. The reward must be something which the member wants and need not be tangible: laughter at a joke is an effective reward, as are other social expressions of support.
- **Coercive power** is the reverse of reward power. It represents the ability to punish or take away something from the member. The punishment must be of significance to the person and may be social rather than tangible.
- **Expert, or informational, power** arises from the knowledge or experience of a person which is salient in the present circumstances. To be effective the group needs to recognize the importance of this expertise to the matter in hand and the legitimate claim of the 'expert' to this mastery. It is inherent in

a person rather than being associated with an organization.

- Referent, or charismatic, power comes from characteristics of the behaviour or personality of a person. It is solely a personal characteristic.

Within the context of a meeting analysing a problem or opportunity it is desirable that expert power form a strong base: the aim of the meeting is to gain insights and the work of the group will be facilitated by everyone contributing their special knowledge. Satisfaction for the individual should be gained from the recognition afforded by other group members. This is more likely to happen if the other power bases are to some degree suppressed.

Perhaps the simplest way to explain this is to consider a peer group: a group, in other words, whose members are equal in major characteristics such as position in the organization. Legitimate power is no longer a concern because no one will have authority, by virtue of position, over another. Reward and coercive power will also be weakened as bases because tangible rewards tend to be linked with authority. The social aspects of reward and coercion will, however, remain and will need to be carefully observed by the analyst. If approbation or denunciation become too intense the attention of the group needs to be drawn to this in order that it can decide how to deal with it. Often all that is necessary to ameliorate matters is to recognize that social reward or coercion is taking place. Referent power may be significant: because the response to it is so highly personal its effect is the most difficult to predict and, ultimately, it is the one with which the group may experience the greatest problem in coping. If the exercise of referent power is seen by the analyst to be having a deleterious effect on the work of the group, a coping strategy is to suspend group meetings and work with members individually, or a few at a time, so that they are temporarily shielded from the influence of the charismatic individual. Full meetings can be resumed after individual members have gained, or regained, some confidence in their own expertise, sufficient to balance the allure of the charismatic.

In groups which have a mixture of members from various levels of an organization, legitimate power and the tangible aspects of reward and coercive power need not necessarily cause problems. If the members are from different sections of the organization, with no members accompanied by people who have direct authority over them, expertise can still form the predominant power base.

The major difficulty comes with a group whose members include staff and managers in direct line relationships. The recognition of legitimate authority will always be a controlling factor, as will reward and coercive power. The extent of the control will depend largely on the hierarchical difference in levels of authority between the members: the greater the difference, the more dominated the 'lower' group will be. It is best to discourage the setting up of a group of this type for problem solving or exploration of opportunities, and to advise instead on the establishment of smaller peer groups with findings coordinated through the presentation of reports. The analyst must be on guard against the possibility of being used as a 'go-between', or reporting link, between the groups: the findings are the responsibility of each group and the analyst cannot speak on their behalf without becoming involved with the ownership of the problem or opportunity.

The individual and the group

Whenever an individual joins a group, or a new group is formed, the contribution of new individuals will be guarded, or filtered, until they are sure of their role and clear acceptance has been signalled through the use of social rewards such as paying attention to their comments, soliciting opinion from them and other 'accepting' behaviour.

The guarding, or filtering, arises from several interacting aspects of perception. In the first place, self-image has a major influence on both what and how something is to be communicated to the group. A person with low self-esteem may be predisposed to say little, and that in a tentative way; the person with strong self-regard may be forthright and full in their explanation. It is important for the analyst to remember that it is the content, and not the style, which will contribute to the solution of the problem. The tentative contributor may need to be encouraged by additional questions whilst the information supplied by the confident person may need to be tested to make sure that no exaggeration has taken place. As a side issue, the style adopted may also give some indication of the need for closer monitoring, reinforcement and support during times when the individuals are working alone on aspects of the problem or opportunity: otherwise, the confident members may overreach themselves and the diffident may under-perform.

The issues of self-identity and making a choice of role acceptable to both the new member and the group are vital considerations and should not be left entirely to chance. If left unresolved the result will be that the new member gradually becomes frustrated, inward-looking and withdrawn from concern with the work of the group. At the same time, no one can design, or force, a role on a new member; what the analyst must ensure is that the new member knows the reason for the invitation to join, what work is expected, the pattern of recent meetings and what work has already been done. Some of this information will be available through the reading of briefing documents, minutes and reports but these will rarely convey much impression of how the group behaves: at least some idea of this can be supplied by the analyst.

The image which the new member has of the other members is also a factor which will cause filtering. Differences in position in the hierarchy of the organization between the new member and other members will be significant. If the new member is lower in the hierarchy this may inhibit the expression of views which might appear to be negative; if the new member is in a higher position than the rest of the group there may be an assumption (perhaps on both sides) that the new member should lead the group. Until the status and role of other group members is clear and the question of leadership resolved a new member will tend to be very guarded. For this reason it is always good practice for the analyst to make sure that new members are aware of the purpose and membership of the group before their first meeting. This gives an opportunity to explain positions and roles in the group and for the new member to prepare for the meeting. In part this preparation is one of self-examination, the new member considering what contribution to make, but it is also a preliminary exploration of the situation represented by the group and the problem or opportunity being studied. Inevitably this will lead the new member to have expectations and these will be clarified, confirmed or rejected as a result of further experience. The analyst can assist the new member to prepare by trying to ensure that the expectations which the member may have are both realistic and sustainable. For this to happen successfully the analyst must also be aware of, and in touch with, the norms of behaviour and roles played by existing group members.

The other group members will also have their own expectations of the new member and their own sense of image, purpose and motives will influence their individual responses. If the project continues over a long time and many meetings it may be helpful periodically to spend all or part of a meeting reviewing progress and discussing feelings. This is especially necessary if the analyst senses that there is ill-feeling or a build-up of dislike amongst the participants. Too much emphasis on the task, without some consideration of the process followed by the group, will lead to the concealment of motives, feelings and intentions. Such issues tend to surface again later on in a project and may lead to delays and a lack of cooperation.

Stereotyping

A particular problem of perception which may be encountered in either individuals or groups is stereotyping. This is one of the ways in which individuals or groups deal with their surroundings: they assign people, events, places, organizations and things to categories and then respond to the category in a particular way. They are responding not on the basis of a particular circumstance but on the basis of a category.

Stereotyping can also be a problem of perception between group members. If group member A has been perceived by group member B as consistently concerned about possible budget overruns then whatever A says about the project in the future may be assumed by B to be about its financial aspects. B no longer listens to A but responds, instead, to the stereotype. Whilst it is impractical to try to prevent stereotypes being formed it is important for the analyst to be alert to the possibility that group members are not listening to what others have to say and should draw attention to this fact. Careful use by the analyst of summaries after each member has made a contribution will help group members to listen to what has really been said.

The individual and the new group: a problem of identities

In the early life of a group the focus of each member should gradually move from self to a group identity. Where this is not tending to happen, members will react in ways which limit their engagement with and acceptance of other group members. There are certain signs of which the analyst should be aware.

The first type of self-centred response is its apparent opposite: a tough, aggressive approach often signalled by ridicule of others, provoking arguments and challenges to whatever leadership or legitimate authority may be present. In essence the person is engaged in a fight which preserves a distance between that person and other members. This 'coping' strategy needs to be distinguished from genuine differences of opinion: in the latter case, the clue may be that the views are expressed with greater consistency and clarity.

Another coping strategy is to seek mutual support, to form alliances and to reduce or avoid opportunities for conflict. Instead of challenges to authority there will be acquiescence and an increasing sense of dependency. Decisions will be avoided or delayed pending further information. Although the output of the group may appear to be high in the sense that reports will be produced and discussions held, little real progress will be made because critical issues are referred to other groups or individuals. The onset of this period of calm may be of considerable relief but it is indicative of a group moving towards passivity rather than constructive problem solving. In genuinely constructive discussions, although the emotional content of a meeting may be low, it will be complemented by a high level of exploration of ideas. It is important not to allow a group to lapse into avoidance as a main strategy; an analyst should make careful notes of what decisions have been taken and, in particular, what action is consequent on them. An occasional reminder of the objectives of the project and a brief analysis of the extent to which they are being achieved by the decisions made and actions proposed is an effective means of modifying the passive response.

Another response for members of a new group is that of withdrawal or denial. Ideas will be rejected by lack of discussion: there will simply be no engagement with the topic. This avoidance strategy can be mistaken for agreement: silence is often misinterpreted as connoting acceptance. The analyst may have to try to break through this response by asking all group members to outline their ostensible agendas and objectives for the project. The task of the analyst is then quickly to identify any contradictions or mutual inconsistencies and to draw the attention of the group to these problems. By reminding the group members of what they are supposed to be trying to achieve the analyst is promoting a culture of greater engagement. This needs to be done firmly, but with tact: if members of the group begin to feel threatened then an aggressive response may be engendered.

The role of the analyst is, at this point, in facilitating feedback to the group on how effective they are being in tackling the workload. This feedback is an important part of building and maintaining a supportive culture but it is also an activity which needs careful preparation. Direct feedback, where the analyst tells members of the group how successful they have been, is rarely helpful because it encourages a view of the analyst as being a leader, and able to reward or punish. Rather, the group needs to be encouraged to discuss how it tackles particular issues; it is at this point that the analyst can provide feedback by offering examples or prompting members to express their feelings.

Analysts are just as much prey to the problem of filtering as any other group member and should, from time to time, reflect upon their own motives and responses to the work of the group. The stance of involvement in the process whilst preserving a sense of distance from the problem or opportunity is difficult to maintain and some conscious effort may be necessary in order to preserve this role.

Signalling agreement

It is often surprisingly difficult for an outsider to discover what decisions a group has made; even more disconcerting is to find that members of the group may themselves not be sure of the outcome of their discussions. The problem often results from a group not having considered how to make decisions and the difficulties which may occur as a consequence. Group decision making is characterised by several approaches, each of which has its place and its consequences, of which the group needs to be aware.

Decision by lack of response is a common though often unacknowledged procedure. Ideas are contributed and then tacitly discarded through lack of discussion. With this kind of decision-making culture an idea has to have a 'champion' and the more power this person has, be it expert, legitimate or charismatic the less likely that it will pass undiscussed. The consequence is that potentially good ideas may be lost because the person presenting them is perceived to lack power. The analyst has an important role to play in noting all ideas and bringing them to the attention of the group until a firm decision is made about them.

Another concealed approach to signalling agreement is for a minority to make the decision. The idea is adopted by a few members whilst the rest remain silent and are deemed to have assented. The lack of response may be cloaking resentment, disagreement or lack of understanding. In many cases it may be unclear to members, other than those of the minority party, that a decision has been taken at all. The analyst should underline what appears to be tacit approval by leading the group back to the point of decision and asking all members to confirm their approval.

In groups where there is a mixture of hierarchical levels, the group may adopt, or the senior person may impose, decision by authority. The senior person takes the decision influenced, perhaps, by the discussions and opinions of group members. This can be an efficient way of making decisions, especially if there is little time for extensive discussions and testing for agreement; it is not, however, a method which is necessarily effective because the other members of the group may feel uninvolved with the implementation of the decision and its consequences. An additional problem is that whilst praise for success may accrue to the leader, blame for failure will tend to suffuse the group members. Not unnaturally, this will create resentment. Unless there is very good reason, such as urgency, for adopting this approach to decision making in relation to projects, it is best avoided.

Decision by majority, using voting and/or polling, is a highly visible test. It has the appearance of objectivity but can result in the losing minority feeling misunderstood and resentful. Its use may provoke reprisals from the defeated group in the form of lack of cooperation or 'spoiling' tactics, applied quite independently of the intrinsic value of subsequent ideas. The truth or validity of the issue on which a vote is taken is not, of course, indicated by the result: the vote is simply a reflection of the degree to which the arguments presented, or the personalities of the disputants, have impinged upon the participants. The winners meanwhile may behave as if truth has been established and their view has been vindicated. Voting is associated with formality and requires the exercise of authority by a group leader. This, in turn, may add to the apparent weight of the decision. Unless necessary to resolve irreconcilable differences or matters of principle, voting is best avoided in the context of soft systems analysis.

Decision by consensus is highly effective for building a sense of cohesion in a group; it can also be very time consuming. In essence it represents a point at which each group member feels that there has been sufficient time for discussion of all points of view, including their own; but it does not necessarily connote unanimity. Whilst most members accede to the view emerging there may also be a minority view: the members holding this should feel that they have had a fair opportunity to make their case and that it has been understood. They should also appreciate why another view has prevailed. In that there has been a full opportunity for disclosure, discussion and persuasion, this approach to decision making in the context of soft systems analysis is to be favoured. The decision will be clear and, provided the discussion has been sufficiently searching, the implications understood. There will be a commitment to implementation even from the minority because there is full understanding about why a particular view has been selected.

The final type of decision-making approach is that of decision by unanimous consent: discussion continues until all opposition has been converted and other views examined and discarded. There is no particular virtue in this approach, despite the long tradition of trial by jury, because the apparent unanimity may cloak unwitting or even deliberate compromises in order to limit further discussion or to appease peer pressure.

Working with the group

It is clear by now that the job of the analyst in facilitating the work of the group is by no means easy and calls for considerable interpersonal skills if it is to be carried out successfully. Much of the foregoing sections forms a shell for the main task of the analyst: ensuring that the problem or opportunity to which the soft systems method of analysis is to be applied receives proper attention.

Solving problems and exploring opportunities involves several different and interrelated activities. They may be divided into two main types: those associated with the task proper and those concerned with the maintenance of relations within the group. Both types are the concern of the analyst as facilitator though, as will be seen, they are not necessarily initiated by the analyst.

Task functions include that of 'initiating', where a member, who may be a leader or could be the analyst, proposes the function of the group and a method of working. The objectives of the group may already be clear from decisions made in other groups or it may be for the group to take a general idea and then decide on which particular aspects to address. A natural consequence of initiation is for the views of other members to be gathered in the form of opinions. This needs careful distinction from the process of gathering information and facts about something: an opinion is not information though the fact that someone holds a particular opinion clearly is information. The gathering of opinion and information both need time, and some judgement must be made on how extensive this process will be. Because the quality of the response of the group depends to a great extent on this stage, the analyst needs to encourage a discussion of how accurate the results have to be. Any decision is a risk because not all factors can be foreseen. The decision about how much information is enough depends on the level of risk with which the group feels comfortable.

Associated with opinion seeking and information gathering are the functions of clarifying and elaborating. Both are limited by the time available for conducting an investigation and by the levels of risk which have been considered acceptable. The analyst can provide useful guidance as to the clarity of ideas and the need for additional information; it is often easy for points of detail to be missed by group members in the enthusiasm of discussion.

Another important job for the analyst is regularly to summarize the discussion in order that ideas are not lost. This also provides a breathing space in which the group can review its progress and remind itself of its direction. Having summarized and checked that the summary is accepted by group members, the analyst has an opportunity to test for agreement or an impetus towards a decision. The result will often give a good guide as to whether further clarification or elaboration is needed.

The second category of function carried out by the analyst is often called 'maintenance'. During the life of the project the analyst must be concerned with the survival and increased capability of the group. Human behaviour being what it is, there are bound to be occasions when people become upset and relationships are temporarily sundered. During such periods the wounded parties will become inward looking and tend not to be favourably disposed to the work of the group. The analyst must seek an early resolution of such problems and ensure that healing takes place. This does not necessarily mean compromising or trying to smooth over differences: it is usually healthy for a group to acknowledge that there are different opinions and it is highly desirable that members see that they will not be punished for holding minority views. One of the most important tasks is to provide a path for the injured to resume their place: the analyst may purposely turn the discussion towards a topic favourable to them and thus encourage renewed participation.

There can be little doubt that working with soft systems analysis requires a good knowledge of technique. Working with and on behalf of a group as a facilitator using soft systems analysis also requires excellent interpersonal skills. Practice and experience are the main ways of gaining such skills but a reading of practitioners experienced in interpersonal behaviour, such as Edgar Schein,[4] is also invaluable if combined with an ability for self analysis and reflection, and a willingness to learn from one's critics.

Terms of reference

External consultants are usually paid for their services but the work of an internal consultant may either be undertaken without direct charge or may be recognized by an internal transfer of funds from one department to another. The same may be true for a group member acting as analyst. The absence of a formal financial arrangement has side-effects, however, because it may appear unnecessary to prepare proper terms of reference for the analyst in relation to the project. An external consultant will expect to enter into a contractual arrangement and it is to the benefit of both parties if the expectations are clear and formally agreed; similarly, for the internally-appointed consultant and the member of a group acting in this capacity there is good sense in documenting what is to be expected from the arrangement. Relying on custom and practice, innate good sense and a feeling of responsibility is often quite satisfactory but will provide little benefit if there are questions within the organization about the conduct of the analysis, the behaviour of participants or the duties and responsibilities of the consultant. The terms of reference provide a set of guidelines for both the consultant and the other members of the organization; the consultant has a difficult role to play and may need the guidance, assurance and authority which is implicit in formally-agreed terms of reference.

Terms of reference provide an agreed statement about the duties, responsibilities and behaviour of the consultant, covering:

- **scope of the analysis** (what parts of the organization, functions or projects will be studied)
- **objectives** (what problem or opportunity is being studied; what questions the analysis will aim to answer. Necessarily, such objectives will be preliminary statements only, representative of the current view of the problem or opportunity)
- **constraints** (areas, topics, problems and people which are excluded from the study; time allowed for the study and financial constraints. Again, these will be preliminary statements subject to formal revision as the study proceeds)
- **resources available** (people and facilities: an external consultant will need some desk space, at least, whilst interviewing, studying documents or working with people; its provision is often haphazard and having to ask for it is not a good introduction to the organization)
- **conduct** (procedures for arranging interviews and meetings, reporting findings, accounting for the use of resources).

A formal agreement covering these points will also serve as an introduction for the consultant to the organization and can also be used to alert other staff to what is going on and how they might be involved. For an external consultant there will also be an accompanying contract covering payment of fees and a formal statement of responsibilities.

Preparing terms of reference may seem an irksome task but it is also one which signals intention and commitment, feelings which are an important part of the process of change.

Keeping records

Throughout an episode of soft systems analysis a lot of information will be requested and a considerable amount will also be gathered by the analyst. In addition, every meeting or interview is a potential source of important information and an opportunity for clarification and verification of information already gathered. Soft systems analysis is not bound by many rules or standards so it is important for an analyst to experiment with different methods of record keeping and, having found a method that works, to establish good habits about using it. To be helpful to clients the analyst must be able to detect inconsistencies, points of agreement and disagreement, and gaps in knowledge; systematic record keeping is important but it must not intrude to the extent that it breaks the flow of discussion during a meeting or interview. Tape recording is sometimes helpful but carries with it the need to transcribe, index, or otherwise abstract the contents of the recording. This can be a tiresome business for a long meeting, especially if it has been volatile. The sheer difficulty of disentangling voices during interruptions is one obstacle; another is the problem encountered with a large meeting where it may be difficult to hear and identify all voices. Meeting rooms are rarely equipped with suitable apparatus and are not suitable as recording studios. In a one-to-one interview the recorder can be placed to pick up both voices with equal clarity, but even here I find it easier to take notes.

There is one cardinal rule which must be applied if tape or video recording is being considered: the participants must be informed of this and give their agreement. Any use of covert recording is unethical and, if detected, will immediately reduce the trust of the participants in the analyst.

During the first stage of soft systems analysis, which is concerned with problem or opportunity perception, it is vital for the analyst to gather all the information about the issue which is available from written and personal sources. The place to start is usually with a written outline or brief prepared by the professed owner of the problem or opportunity. Even if the owner and analyst are well known to each other and confident in their knowledge of the background and circumstances, commitment expressed through the act of writing is an important first step in encouraging an analytic frame of mind.

If the analyst does not know the background of the organization then information about this should be requested. What information is needed will depend, to some extent, on the type of problem or opportunity being explored. There is a basic minimum that should be made known at the outset, however, and without which it will be difficult for an analyst to work:

- organogram showing positions, responsibilities and reporting lines
- recent annual reports, financial reports and statistical digests relating to the work of the past three years
- mission statement and strategic plan showing objectives
- list of current staff showing names, locations, telephone and other contact details (an internal telephone directory is often the best source)
- if the organization is dispersed, locations of branches and headquarters with brief details of the function of each branch, together with staffing arrangements if these cannot be deduced from the organogram
- brief details of any legislation, standards or other validation by external bodies which circumscribes the activities of the organization.

This information provides a profile of the organization and allows the analyst to form some tentative idea about what the organization thinks it is doing and what it seems to be doing. The difference between these perceptions is often significant and can be a fruitful starting point for discussions.

The information gathered at this point of the investigation will be a mixture of written items, records of discussions, ideas and further questions. It will be difficult to avoid the feeling of being surrounded by complexity and almost overwhelmed by the volume of information. Careful records will have to be kept but the analyst needs to remember that what is needed at this stage is a broad understanding of how the organization and the people within it perceive the problem or opportunity.

How notes and records are kept depends on the preferences of the analyst. Some favour extensive written summaries of each interview, meeting or document, with references to associated documents; others prefer to capture the information using a technique known as 'conceptual mapping' to produce a diagrammatic summary; more will said about diagrams and mapping in Chapter 6.

A simple technique, which I have found helpful, is to allocate a reference number to every document I am given and every document, such as a summary of a conversation, I produce. The reference consists of a date followed by a running number for the document; for example, the third document encountered on the 13th April 1996 will be given the reference 13/04/96:03. The documents are kept together in strict chronological order in a box file. I also prepare a summary list showing a brief title for each document and its reference number. In the early stages of a project there may be no very clear structure to the investigation but, gradually, the main components become clearer and I can tentatively begin to assign documents to a structure. 'Staff implications', 'costs' and 'technology' are headings that often crop up. For each component I prepare a summary sheet listing the relevant documents and their reference numbers, what is known about the component and what questions or uncertainties remain. A document may appear on more than one summary sheet; no document is, however, removed from its place in the chronological file. This file remains at the heart of the study and it is surprising how often its sequence assists me to locate a particular item, especially if it is used with my other basic tool, a diary showing the record of meetings and participants in interviews. Core documents, to which frequent reference is made, can easily be distinguished with coloured adhesive tags for quick location. 'Rich pictures' used in the course of discussions are copied from chart paper onto A4 size paper for inclusion in the file; although they are essentially working documents, used to promote and record discussion, I find they are also a powerful means of recapturing that discussion.

So far I have been unable to find suitable software for the task of keeping such records on a computer. Although I use a word-processing package for most writing, and find it almost indispensable, the task of setting up a database, reconfiguring it as my ideas of the project change, and entering all the necessary details is not attractive. In addition, I often find it helpful to browse in the summary sheets and this seems most easily done if they are kept in paper form. The craft of analysis depends on being open to new ideas, facts and opinions. Not all of these come from purposive investigation but often from the combination of serendipity and a prepared mind. The summary sheets are often useful when working with an individual or group: a computer seems a cumbersome and inappropriate substitute on such occasions. This is an applications area, however, where new software is being developed and new ideas tried. Provided the cost of purchase and the investment of time to learn to use a package satisfactorily can be justified, there is nothing intrinsically against the use of information technology in the conduct of soft systems analysis.

References

1   Ivancevich, J. M. and Matteson, M. T., *Organizational behavior and management*, 4th edn, Homewood, Irwin, 1995.

2   Bennett, R., *Organizational behaviour*, London, Pitman, 1991.

3   Turton, R., *Behaviour in a business context*, London, Chapman and Hall, 1991.

4   Schein, E. H., *Process consultation: its role in organization development*, Reading, Massachusetts, Addison-Wesley, 1969.

Chapter 4

Soft systems analysis

I N THE PRECEDING chapters the distinction between hard and soft systems analysis has been discussed and the role of both within the framework of planning defined. This and the next two chapters form the core of the book: they aim to give a comprehensive account of soft systems analysis and of how it can best be conducted. A model of the analytic method will be explored from a theoretical point of view and its practical application discussed. The model will be investigated sequentially but it is important to recognize that soft systems analysis is a very flexible method which does not depend for its success on a rigid adherence to this sequence. In other words, although there is a strong underlying model for soft systems analysis, it is not one which enforces a series of actions or procedures on the analyst. When first beginning to use the method it is good sense to follow the sequence suggested by the sequential model because this will ensure that a significant step is not omitted, but it will quickly become apparent to the more experienced analyst that one can start the analysis at various points of the sequence and select the particular analytic view which will suit a problem. If, at a later point in the analysis, a view from earlier in the sequence is needed, it is easy to run the earlier stage which would yield that view and then combine the results with those from later stages.

A model of soft systems analysis

Figure 4.1 presents an outline of soft systems analysis. At first sight it may appear unlike the model of planning explored in Chapter 2, although some similarities will become evident on closer examination. Definition of needs and choice of implementation strategies are steps included in both. The differences are in the use of conceptual

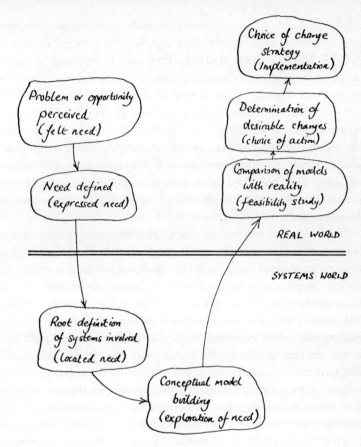

Fig. 4.1 *Outline of soft systems analysis*

model building, comparison of the models with reality and in the establishment of a root definition. It is these features which provide the distinctive characteristic of the soft systems method.

There is another point of contact which may be helpful to practitioners of library and information science: the process exemplified in the soft systems model bears some similarity to the techniques used in the reference process. The phrases in parentheses included in Figure 4.1 are intended to highlight this point. The reference process starts with a person experiencing an information need (felt need). To satisfy this need the person may eventually decide to ask for help from an information worker: in doing this, the person becomes a client of the information worker and the need is expressed in a spoken or written form. In discussing the reference process, the importance of this stage is rightly stressed: if the client does not correctly and fully convey the essence of the information need an effective and satisfying solution is unlikely to be reached. The 'reference interview' centres on encouraging the client to discuss the need, arrive at an agreed definition and then to explore it in order for the information worker to understand its significance and ramifications. The information worker can then decide which sources should be consulted to provide a solution.

The reference process depends crucially on communication between the client and the information worker. Problems in this process can severely compromise the likelihood of success in satisfying the information need. In this respect, soft systems analysis and the reference process are analogous. They differ, however, in the knowledge base used. For the information worker the knowledge base consists of an understanding of the reference process itself and of the contents of information sources and how they may be used. The information worker will also need some understanding of the subject of the information need of the client. For the soft systems analyst the knowledge base consists of an understanding of the process of enquiry and how to assist clients in formulating statements of need. Although an understanding of any matters associated with the need of the client may be helpful, it is not a necessary qualification for the analyst to be able to assist the client. The skill of the soft systems analyst is in working with the client on the *process* of enquiry rather than the technical solution to the problem.

There are, indeed, points at which knowledge of the technical component of the client's problem can become an obstacle in the process of soft systems analysis. Between people with a shared understanding of a technology there is often a recourse to jargon and an unvoiced expectation of shared assumptions about the nature of the problem they are discussing. Where people have been working closely together for some time and have experiences of the technology in common, this development is to be expected and can be beneficial. It allows discussion about complex topics without a constant need for definition and exploration of topics over which there has already been agreement. A convention based on this exchange of experience and understanding is established and works well, so long as the problems being explored are firmly rooted within the boundaries of this area of shared belief. The knowledge base of the workers in an organization forms a most valuable component of the available information resources.

Many of the problems which can most fruitfully be studied through soft systems analysis are near, or beyond, the boundaries of any such shared belief. A client may decide to seek assistance with a problem because the resources inherent in the knowledge base of the workers and the organization seem inadequate. If the analyst displays an obvious technical knowledge recognizable to the client then there will be a temptation to relax into the use of jargon and the assumption of knowledge of shared experiences. The critical role of the analyst is in encouraging the client to subject all assumptions relevant to the problem and, perhaps, some which do not at first sight seem relevant, to a searching scrutiny so that the client develops a clearer understanding of the problem and its context. It is not the analyst who is developing a solution for the problem: rather it is the client who is gaining a fresh view of the problem in order to develop a solution.

Another aspect of the model is that, unlike the planning model, it is not cyclic: the end point represented by the 'Choice of change strategy' does not lead neatly back into the beginning of the model. This is because soft systems analysis is only one component in a planning system: it is not, of itself, a method of planning. The results of the analysis could lead into the application of hard systems analysis, if the problem analysis identifies some application of technology as being appropriate for meeting the need. The analysis might reveal, for instance, that the problem is attributable to policy decisions rather than involving strategic or administrative procedures; it might suggest the need for an assessment of staff morale or job satisfaction; it might draw attention to a need for evaluation of the effectiveness of some services. In each case the 'Choice of change strategy' will consist of defining particular action steps which may include specialist analysis requiring expert knowledge (knowledge of information technology, personnel administration or policy formulation, for example). It is out of these action steps that a solution to satisfying the need will be developed. The end of an application of soft systems analysis leads on to other techniques rather than feeding back into itself.

It is also important to realize that any process of analysis inevitably changes the nature of a problem because new knowledge and understanding emerge. The choice of a loop to model this would be inappropriate because a cyclic process encourages an assumption that the context of the problem being explored remains substantially the same at the end of a cycle as at the beginning. Whilst this may be true in some instances, a safer course is to examine the context afresh. Such an approach will ensure that the participants are prompted to share new knowledge and insights as well as agreeing on the context. In addition, it must be remembered that the participants will have been affected by the passage of time, their own experiences of previous work on the problem and, it is to be hoped, their enhanced understanding of it. Unfortunately, unless a positive step to recapitulate and explore the context is made, the tendency is to skip this and move directly on to analysis.

The model also shows a move from the 'real world' into the 'systems world' and back. At first sight this distinction may alarm new practitioners, and those for whom they are conducting an analysis, because it suggests a sudden retreat into areas of theory and abstraction, with little apparent connection to practical problems. Whilst the division does indeed indicate a move by the analyst into systems-centred thinking, it is followed by a return to check that the insights gained during this stage are also applicable and valid in the 'real world'. The move should not be seen as one of withdrawal; rather it is a shift of focus which may be undertaken in the absence of the client or, if preferred, with the active cooperation of the client.

A final aspect of the model, mentioned earlier, is worth emphasizing: although it represents a chain of activity, the method does not rely upon each stage being completed before the next is begun; nor does it depend on each stage being used or applied in a set sequence.

The stages can be thought of as tools in the toolbox of soft systems analysis: each has a use and, for particular jobs, some need to be used before others; not every tool will be required for every job.

Having looked at the model as a whole, a more detailed study of its components can be undertaken. Notice that this is an example of a move from 'system' to 'subsystem' level: the soft systems model is the system and its stages are individual subsystems, connected together, with each relying on earlier stages for information to process.

Perceiving problems and opportunities

The first stage of the soft systems method, the perception of a problem or opportunity, is one which is likely to be used on most occasions when the method is used. As we have seen, not all problems or opportunities need soft systems analysis: the key determinant is how certain the decision makers are about the nature of the problem or opportunity.

It may seem almost unnecessary in this book to discuss how managers perceive problems and opportunities: surely there can be little mystery about this? However, to use an earlier metaphor, there can be many and varied views of the same landscape. The soft systems method is ideally suited to dealing with circumstances where there are such multiple views and particularly where these views are difficult to express or ill-defined. The study of management often focuses on the orderly use of techniques; Ackoff[1] once described the job of management as dealing with messes, and this is an excellent way of distinguishing between the types of problem which are already in a suitable form for 'hard' approaches and those which are more diffuse and need a 'soft' approach – as a preliminary, at least. To reinforce what was said in an earlier chapter, if the objectives, the means of solving the problem and the criteria for choosing a solution are clear there is nothing to stop immediate use of the hard approach. Few problems or opportunities arrive in this form, however, especially those which have a strong 'people' component in them. As far as libraries, information services and resource centres are concerned people, whether they be staff, users, or other stakeholders, are usually strongly involved and this suggests that soft approaches will often be a useful start to exploring a problem or opportunity.

'Human Activity' and 'Designed' systems

Most writers on soft systems analysis distinguish between 'Human Activity' systems and 'Designed' systems. Human Activity systems encompass people doing something for some purpose, whereas the 'Designed' system connotes something constructed, such as machines or a building, and abstract things such as a classification scheme or a cataloguing code. At first sight it might seem as if soft approaches are applicable to the Human Activity systems and hard to the Designed systems. However, many aspects of work involve the use of things within a Designed system by people in Human Activity systems: in other words, many areas of work depend on the link, or interface, between the two systems. An example drawn from the milieu of libraries, information services and resource centres is that of the contrast between a philosophical classification of knowledge and a bibliographical classification. The former may emphasize the use of logical division and derivation of topics. The latter, to be successful, must be grounded in topics about which documents have been written: there must be literary warrant for the inclusion of a class.

This close link between Human Activity and Designed systems is a distinctive characteristic of many organizations which are principally concerned with providing a service. In such enterprises, the constructs are tools used by staff on behalf of the users and the community rather than being the main focus for the organization. The tools facilitate the provision of the service and amplify the capacity of the staff to assist, but it is the staff who are responsible for providing the service.

Systems and the service culture

The principal focus of libraries, information services and resource centres is the provision of service to users. Even though the user approach is emphasized in mission statements and objectives, the increasing use of technology can sometimes lead their managers to take a narrow view of the service by focusing too exclusively on the role of technology. An example is the design of some early Online Public Access Catalogues (OPACs) which reflected the capabilities of computers rather than the needs of users.

It is also worth considering how features of the Designed system can be misperceived by users and their value not properly assessed. Electronic information technology is a good example of the whole being greater than the sum of its parts, because some users, and perhaps some neophyte information workers, assume that the answer gained from its use must be better than the answer provided by manual means. The use of an advanced technology adds an allure to the service which may not actually be reflected in significantly better results or a more effective service, though the user may feel it to be more substantial. To those who have been involved in the design, and to expert users, the strengths and weaknesses of the approach should become clear and should discourage inappropriate use; it is the newcomer and the infrequent user who may need additional guidance.

It is often not sensible to consider the Designed and the Human Activity systems separately because design should take account of how people want to work, how they will use what has been constructed and how the construction may, in turn, influence how people behave. The link is subtle and is, unfortunately, often forgotten. For this reason, it is beneficial to spend some time in thinking of the constructs as being extensions of the Human Activity system rather than being separate components.

Ownership

Who owns a problem or opportunity? This may seem a strange question to ask because common sense would suggest that it is the person who first becomes aware of the problem or sees the opportunity. This is not an entirely satisfactory answer, however, because the person who detects it may not be able, or willing, to do anything about it other than drawing it to the attention of someone else.

Ownership must instead serve to connote awareness, assumption of responsibility and control of resources suitable for coping with the problem or opportunity. This serves to underline the point made in Chapter 3, that the analyst has a distinct role to play which should not be confused with that of problem ownership.

The definition allows multiple ownership, a form which commonly occurs in organizations using a management team. The problem or opportunity is taken up, by common consent, by the team, which collectively contributes towards a solution or plan. Some care is needed with multiple ownership, however: with a large team it is easy for the concept of ownership to become so diffuse that no one feels accountable. Handing a problem, or the responsibility for exploring an opportunity, over to a smaller team will help to retain a clear sense of responsibility, but only if the larger management team retains some control through the setting of targets and timetables for reports. In essence this represents an elaboration of the notion of ownership, with the smaller team owning the practical stages of coping with the problem or opportunity, whilst the larger retains overall accountability for it. It is important that the facets of ownership remain clear.

An allied difficulty can be that of 'protective' ownership. In these circumstances a group develops a strong loyalty within itself and a strong identification with the project which has been entrusted to it. Whilst this attitude may be admirable for encouraging dedication and hard work on a project, the disadvantage is that the group will seek to protect the project, and itself, from criticism, whether specific or implied. Comments or suggestions will tend to be resented unless they are clearly complimentary. To ameliorate such disadvantageous effects the management team need to hold regular meetings with all or, at least, the main members of the smaller group in order to emphasize their continuing stake in ownership. In addition, the team and the group together should plan a timetable of progress reports.

The stages of soft systems analysis

Although the flexibility of soft systems analysis has been mentioned it is helpful when first encountering and making use of it to go through the various stages in a set sequence. As can be seen from the model (Figure 4.1) there are seven identifiable stages.

Stage 1: Problem or opportunity perceived

Stage 1 is concerned with the perception of the problem or opportunity. Checkland[2] has identified this as being a surprisingly difficult area to negotiate, not least for the problem owner or owners. He suggests that perception of problems and opportunities can range along a spectrum, with one extreme being the declaration that the organization has no problems and has adjusted to suit the changing circumstances it experiences. The other extreme is the organization where the response is to acknowledge that there may be problems but that they are insufficiently understood even to be specified.

Both extremes are characterized by elements of vagueness or uncertainty. The organization that has no problems and cannot, seemingly, be improved is actually in deep trouble because its attitude is reflective of a closed system: such a system is unresponsive to its surroundings and, in the end, will disappear, either because it is no longer valued by stakeholders in the environment or because the environment will destroy it.

Unfortunately, to persuade the members of the organization to spend some time in analysing the objectives of the organization, what they do and what their present and potential users need, may be quite difficult. So strong may be the belief that there is nothing wrong that a 'neurotic loop' may have built up whereby any suggestion that change should occur is neutralized. Consider Figure 4.2: the model starts with a condition of stability, which is perturbed by some pressure for change. This pressure may be exerted through the asking of questions, which would imply the need for change, or some proposal or request. Whatever the nature of the pressure it is perceived as coming from outside the organization or, at least, to bear only upon a smaller group within it.

Change almost always carries connotations of threat. It represents contact with the unknown and the instinctive reaction when the pressure for change is perceived is for a period of disruption to intervene. The disruption is manifest because of the varying perceptions which are occurring. Typically, some people in the organization or

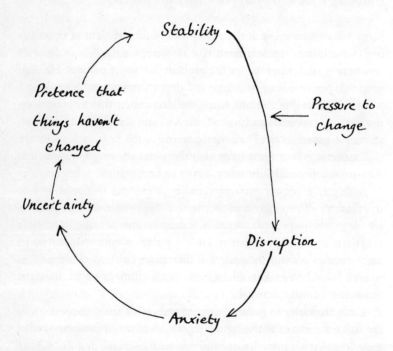

Fig. 4.2   *Neurotic loop*

group will remain insensitive to the pressure and not even acknowledge its existence. For those that do perceive its existence there will be disagreement about its shape and form. No clear view of the pressure to change will emerge and certainly no coherent opinion on how to manage it. Disruption quickly degenerates into an atmosphere of anxiety as individuals consider their own responses. Anxiety reflects the individual's awareness of personal vulnerability. The response to this, by most people, is to seek others with whom to compare views, seek mutual protection and consider what to do. Without intervention and guidance the ensuing uncertainty will lead to a response which serves to reduce the apparent threat by encouraging a pretence that the environment has not changed after all and that there is thus no need for the organization or group to change.

Whilst this model presents an unedifying picture of a group or organization failing to make a positive response to change, it must be accepted that in some circumstances a decision not to change is appropriate. The point about the model, however, is that there has been no decision: instead, there is an avoidance of the issue.

There is one hopeful sign in the model. The period of disruption is one during which there is an outwelling of energy and activity; the problem is that it is undirected and will later be dissipated in anxiety. It is at this point that soft systems analysis can most profitably be used to intervene and serve as a channel for the energy.

If undirected energy is both a problem and a saving feature of the 'neurotic loop' model, the same may not be true at the other extreme of problem or opportunity perception. Here, the recognition that there are problems about which little is really understood may be accompanied by a flaccid, hopeless atmosphere in which it is difficult for the people in the organization or group to raise the energy to carry out further studies. Sporadic attempts may have been made to tackle such problems but, in the absence of a clear sense of direction and some confidence in the possible existence of a solution, such attempts are likely to collapse out of inertia. Whoever is conducting the soft systems analysis will have to work hard to convince the other members that it is worth exploring the problem and that the task of definition, at least, is not beyond the capacity of the group. In part, the despondency may come from not having shared what information and views are available: the isolation felt by each person is a major impediment to creativity. The task of the analyst is to assist the group to find the requisite energy by helping it to gain a clear understanding, an agreed direction and a commitment to developing a solution.

Starting from a position of previous failure or lack of engagement is never easy. The analyst may have a slight advantage if appointed as an external consultant because this will provide a reason for exploring what the rest of the members of the group regard as 'old ground', unpropitious for further work. Even so, the attitude of the group members may be to contribute what they know and then sit back, disengaged and awaiting an answer from the analyst. Although perhaps tempted by such circumstances to seek to provide a solution, the analyst must resist, so as not to assume ownership of the problem and to be able to concentrate on two aspects of the individual presentations: points of agreement and points of disagreement.

It is most unusual to find a group where all the members concur or have completely contradictory opinions from the outset. Wherever there is a shade of opinion, however slight, there is a starting point for discussion. The analyst must seek to raise the consciousness and energy of the group by providing a forum and then drawing attention to apparent facts and opinions which are not completely accepted by all group members.

It is to be hoped that an analyst will not always encounter circumstances at either extreme of problem or opportunity perception, although it must be acknowledged that consultants tend to be appointed when seemingly intractable circumstances have become apparent.

As an example which we can follow through the stages of a soft systems analysis, consider the following problem. The library of a university has, over the last few years, begun to experience quite high rates of turnover of senior paraprofessional staff. Many of the staff have applied for jobs in other types of organization (banks, large shops and so on) but there does not appear to be a big salary differential or major difference in benefits compared to their jobs at the university. The University Librarian has recently adopted a policy of conducting an 'exit' interview with anyone resigning from the staff and has noticed that the paraprofessional staff tend to give their needs for greater job interest as a principal reason for leaving. The University Librarian and the rest of the senior management team are concerned with this increased turnover because there is a shortage of suitable people to promote into these positions and training is a costly exercise. Ideally they would like a person occupying a post at this level to stay for at least five years.

The problem is perceived as one of motivation. The senior management team has explored several aspects of motivation, remembering that increased salary is not necessarily a motivating force for all people in all circumstances and also acknowledging that uncompetitive salaries have rarely been mentioned as a reason for leaving. In any case, changing the salary and benefit structure is not under their control; it could take a long time to prepare a case for presentation to the University Staffing Committee and the result would be very uncertain.

Stage 2: Need defined

As information is gathered during Stage 1 certain ideas and facts will become apparent: some will be certain, others tentative. They will be related to each other and, taken together, will form a preliminary statement of what is known, or thought, about the problem or opportunity. Soft systems analysis makes use of a 'rich picture' to summarize and convey this understanding. A rich picture is, to some extent, like a cognitive map or network of relationships: it expresses relationships and affinities though these are not necessarily confined to hierarchical and associative relationships. The technique of drawing rich pictures is explained in Chapter 6. For each participant, or for subsets of the complete group, there may be a separate rich picture representing a distinctive view. Another approach is for the group to develop a single rich picture but, for this to be successful, there needs to be a high degree of agreement on the characteristics of the problem or opportunity.

The essence of the definition of need, whether this be related to a perceived problem or an opportunity, is to be able to identify the gap between present circumstances and 'what ought to be'.[3] The need cannot be clearly defined until the gap between the two has been recognized. This is not often immediately obvious; indeed, one should be cautious if it does appear to be obvious because this may be a sign that the need is superficially understood or even completely misperceived. Solving the wrong problem will provide all the appearance of progress but it will still produce the right answer to the wrong problem.

What is required is an agreed statement about what the need is; at this stage there will, almost certainly, be different views of the need and its implications but there must be agreement on the statement of need itself. If there is not, this is a sign that discussion must continue in order to reach consensus or that there are multiple needs associated with the problem or opportunity, each of which should be specifically defined. In the case of the apparent existence of multiple needs the analyst must be prepared to accept that not all group members may agree on the multiplicity or the number; each apparent need should have some significant support or, at least, a well argued case for its recognition. In cases of doubt it is best to include a need: often in the later stages of analysis such doubts are resolved or it becomes apparent that the doubtful need can be conflated with another to form an encompassing statement of need.

A satisfactory statement should cover five aspects of definition:

- a succinct account of the problem or opportunity – for example, 'increased mutilation of printed materials'
- the locus of the problem or opportunity – for example, 'in the reference section of the Education Faculty Library'
- the stakeholders who are directly affected – for example, '65 academics and 850 students of the Education Faculty and an estimated 400 external users'
- the scale of the problem or opportunity – for example, 'in the past three months, the number of known cases has doubled, compared with previous periods. Thirty mutilated books have been discovered either by library staff or as a result of complaints from users'

- the impact of the problem or opportunity – for example, 'the reference section is used intensively by students engaged in project work; library staff have received a greatly increased number of complaints from both students and academics over the last three months'

This provides key information with which to judge the seriousness of the issue and the need for a response. There may still be areas of contention or doubt surrounding aspects of the statement and these should be noted for further discussion. To follow the example used above, there may be some dispute about whether the number of acts of mutilation is suddenly increasing or whether the library staff are being more rigorous in their checking of reference stock; what cannot be in dispute is the serious impact on users. An area for further enquiry may be whether a similar apparent increase in acts of mutilation is also evident in other faculty libraries in the university.

The rich pictures and the statement of need provide the main focus for the next stage and their production will, almost invariably, mark a point of transition from the 'real world' into the 'systems' or 'meta-world'. To reinforce a point made in an earlier chapter, this is a change of view undertaken at the instigation of the analyst and should not serve to exclude the other participants in the analysis. It is, however, often beneficial for the analyst to prepare a presentation of the next stages for comment and elaboration rather than to try to develop them directly with the group. Group members other than the analyst may be largely unfamiliar with soft systems analysis and the effort involved in teaching will divert energy from the task of analysis itself.

To return to the case study of the paraprofessional staff in the university library, having perceived the need as being that of motivation of the staff involved, the next step is to explore that perception. The starting point is to discover if there is agreement on the validity of the perception. The views of paraprofessional staff are sought and it is confirmed that salary and benefits are not a major reason for discontent; however, the level of repetitive routine work is a source of complaint. It is noted that the problem of high turnover is not so marked in the Processing sections of the library but is especially apparent in the Reference and Circulation sections. The paraprofessional staff concerned feel that their jobs could be done equally well by staff in less senior positions. There is an apparent need, therefore, to enrich the jobs of the paraprofessional staff, particularly in these two sections. The discussion is captured in the form of a rich picture which reflects an initial understanding of the problem (Figure 4.3).

By now the need is clearly defined in terms of the five aspects of definition outlined above:

- succinct account of the problem – turnover of paraprofessional staff is too high
- locus – Circulation and Reference sections of the University Library
- stakeholders – four paraprofessional staff, two in each section; the Heads of the Circulation and Reference sections; University Librarian

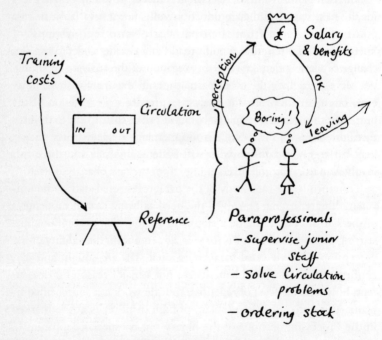

Fig. 4.3   *Rich picture of the 'paraprofessional staff' problem*

- scale of the problem – in the past two years these four posts have been vacant on five occasions
- impact – Training takes about two weeks. Errors have an immediate impact on the quality and efficiency of the service, sometimes directly affecting users

Developing this definition of need provokes a lot of discussion centred on the range of duties undertaken by the paraprofessional staff. Paraprofessional staff in the Circulation section are given training in the use of the computer system to enable them to amend circulation records if they are incorrect, register users and grant privileges. They are also responsible for supervising junior staff and allocating them their duties. Paraprofessionals in the Reference section are given training in supervising the staff who re-shelve the stock, organizing the timetable for five junior staff and in the routines of stock ordering, including bibliographic checking.

The senior management team decides that it would be best to consider the Reference and Circulation sections separately because it is clear that the range of duties currently associated with paraprofessional staff in each section is different. They decide to deal with the Reference section first and commence by briefly reviewing its activities. This is captured in another rich picture (Figure 4.4). What this reveals is that the professional staff of the Reference section deal directly with users and their queries; paraprofessional staff have little contact with users. The process of answering a query starts with a decision by the professional staff of the Reference section as to whether a particular query falls into one of three classes:

- quick-reference – a general knowledge query which can be answered from the stock (excluding off-campus computer services) of the section in less than ten minutes
- extended queries – a query which will require more than ten minutes work but does not require detailed subject knowledge or use of resources outside of the Reference section
- referred queries – queries which require detailed subject knowledge or access to specialized stock or information services and which are referred to the appropriate subject librarian

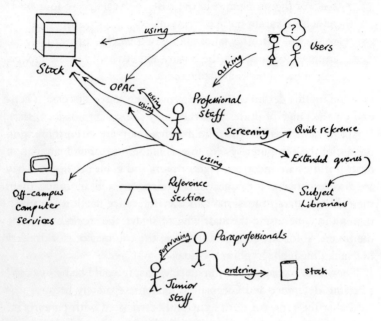

Fig. 4.4   *Rich picture of the Reference section*

One of the suggestions which comes up during the discussion is that some job enrichment for the paraprofessional staff could come about if they were to be involved in the work of answering quick-reference queries. The senior management team agrees that this idea is worth considering.

Stage 3: Root definitions

Until now there has been a tacit acceptance that there may be several different views of a problem or opportunity and that these are part of the reality of exploring problems and opportunities. This acceptance has been signalled by the development of rich pictures but these have also been used to focus attention on the production of an agreed statement of need. The production of rich pictures is part of the process of exploring the project area in order to reduce the areas of disagreement between group members. In the course of a study it is also likely that several systems or aspects of the work of the organization will have been identified as being involved with the problem or opportunity. The final step in this process of coalescence is the development of 'root definitions' of the Human Activity systems which seem relevant to solving the problem or taking advantage of the opportunity.

A root definition represents a choice of view of the problem or opportunity from amongst the several which may have emerged. In order for there to be progress on developing a solution or planning a project there needs to be agreement that certain views seem more promising or more pertinent than others. The function of soft systems analysis is both to facilitate a wider understanding of a problem or opportunity and to highlight those views which seem likely to lead to desired solutions. There needs to be agreement on the purpose and aims of a Human Activity system, and this requires agreement on who and what is affected by what it does.

Continuing with the case study, it has been agreed that it would be worth considering the quick-reference service as a possible area for enriching the jobs of paraprofessional staff. The agreement signals that the quick-reference service has become a Human Activity system worth investigation and study in its own right. The services offered by the Reference section represent, as a totality, a Human Activity system of which the quick-reference service forms part. Now the decision is to move the focus of the study to a subsystem of the Reference section: the quick-reference service.

Six components are necessary to construct a root definition; these are usually called the 'CATWOE' components, a mnemonic derived from the first names attributed to the components. To illustrate the components, the case study commenced in the previous sections is again used. The components, in mnemonic order, are:

- **Clients**: *those who benefit from what the system produces*. In discussion it becomes clear that the quick-reference service is offered to students and members of the academic staff of the faculties of the university. Students and members of faculty therefore benefit as clients of a quick-reference service system.

- **Actors**: *those who carry out the work within the system.* Professionally qualified staff of the Reference Library would be some of the Actors involved in the quick-reference service system.
- **Transformation**: *the change which the system causes to happen; the process by which an input is changed to a particular output.* The quick-reference service system aims to transform clients with an information problem into clients with an answer to that problem.
- **Worldview** (sometimes referred to as 'Weltanschauung'): *the perception of the system from a distinct point of view; the assumptions which are made about the system from that perspective.* The Worldview of the quick-reference service system is that it is a system for supplying answers to problems which need less than ten minutes research.
- **Owner**: *the person or organization who has ultimate authority over the system, who can cause it to cease to exist.* The Owner of the quick-reference service is the Reference Librarian.
- **Environment**: *what surrounds, or lies outside, the system; the system may influence it but has no control over it.* For the quick-reference service system, the Environment is the stock of the Reference Library, excluding use of off-campus computer services.

The components of a root definition are then used to compose a succinct description of the Human Activity system, with each component being reflected in the description. Checkland has suggested a general form for such descriptions:

A (. . . O . . .)-owned system which, under the following environmental constraints which it takes as given: (. . . E . . .), transforms this input (. . .) into this output (. . .) by means of the following major activities among others: (. . . . .), the transformation being carried out by these actors: (. . . A . . .) and directly affecting the following beneficiaries and/or victims (. . . C . . .). The world-image which makes this transformation meaningful contains at least the following elements among others: (. . . W . . .).4

For the example of the quick-reference service, used above, the description would be:

> A system owned by the Reference Librarian within which professionally qualified Reference Library staff, using the stock of the Reference Library but excluding use of off-campus computer services, seek to produce answers to information problems from students and members of faculty which take less than ten minutes research.

The elements are often gathered together into a small picture (Figure 4.5) as a visual aid for discussion.

By taking other points of view, it is possible to develop other root definitions. The paraprofessional staff might, for example, see the quick-reference system as a device for reducing the intrinsic interest of their jobs by denying them an opportunity to seek answers to some of the questions; subject librarians might see it as a useful means for enabling them to concentrate on the more difficult questions asked by faculty and students.

Brief consideration of these alternative perspectives highlights the point that the core of the CATWOE approach to developing a definition is the link between the Transformation process and the Worldview. Taking a different perspective can offer insights into how the process of Transformation can be achieved in other ways which may be of greater benefit to the Actors, Owner, or Clients, or may alter the influence of the Environment. In-service training of paraprofessional staff to answer defined types of information problem, such as transport timetable enquiries, or the use of Internet connections to sources of information other than the Reference Library collection, would be ways of altering the system whilst still retaining the same Transformation effect.

The value of the root definition approach is that it encourages a specific consideration and statement about who is involved, and how, in a particular task or set of tasks. The process of discussion may reveal misunderstandings, hidden constraints and unexpected views. It also helps to ensure that relevant information has not been neglected or overlooked. Above all it encourages thought about the

Root definition : a system owned by the Reference
Librarian, within which professionally-qualified
Reference library staff, using the stock of the
Reference Library, but excluding use of off-campus
services, seek to produce answers to information
problems from students and members of faculty,
which take less than ten minutes research.

C — students and members of faculty
A — professionally-qualified Reference library staff
T — information problems → answered problems
W — information problem must need less than ten minutes
    research by staff
O — Reference Librarian
E — resources of the Reference library

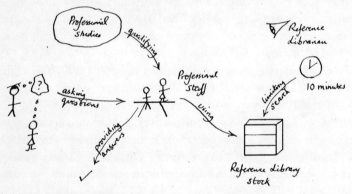

Fig. 4.5   *Root definition of the 'quick-reference' service*

design of the Human Activity system in order to improve it.
Understanding the Human Activity systems which are the milieu for
the problem or opportunity being considered is an essential step in
resolving what to do.

Stage 4: Building a conceptual model

The conceptual model of a system takes the Transformation already defined in the root definition and then prompts identification of the minimum set of activities which must be present to carry it out. Because it is based on activities, the model is labelled with verbs and nouns which represent each activity; lines and arrows are used to link activities. The conceptual model is an extension of the root definition but should still represent the same level as that of the root definition (Figure 4.6).

The conceptual model shows a typical 'question-answering' sequence, familiar to anyone who has studied the reference process. What triggers the sequence is the arrival of a faculty member or stu-

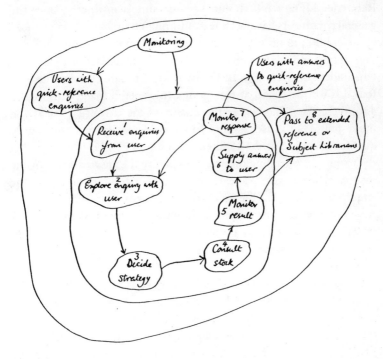

Fig. 4.6   *Conceptual model of the 'quick-reference' service*

dent (the 'user') with a quick-reference enquiry. This underlines an important point about conceptual models: there is usually a set of procedures to monitor the process of transformation and then to take action depending on the results.

In the case of the quick-reference system there has to be a monitoring process present in the Reference Library which decides whether a question can be answered by use of the quick-reference approach or whether it should be routed to extended reference or the appropriate subject librarian. Notice that the process of monitoring may be intrinsic or extrinsic, depending on the formality and methods used. Job evaluation is an example of an extrinsic monitoring process which is formal and depends on a set group of methods. Making a decision about the type of reference path to pursue is an intrinsic part of the job of the professionally qualified staff of the Reference Library and should become almost instinctive once the general parameters have been learnt through training and experience.

In addition to prior monitoring and control there is usually some monitoring and control process to be found associated with the system under consideration. In the context of the quick-reference system this can be found at two points (numbered 5 and 7 in Figure 4.6); these are points where the quality of the answer (point 5) and the response of the user (point 7) are considered by the professionally qualified member of the Reference Library staff who is undertaking the enquiry. Depending on how satisfactory the answer seems or the response of the user appears, the member of staff may pass the enquiry on and out of the quick-reference system. In addition, at the stage of monitoring the response from the user (point 7) the member of staff may decide to try the quick-reference route again if the user appears not to be satisfied.

A system boundary, shown on the conceptual model by an encompassing line, is also added at this stage. The process of analysis moves from a hazy idea of the components and perimeter of the systems under investigation and gains delineation gradually through the work in the 'systems world' where aspects such as boundaries and structural elements (for example rules and procedures) are formally considered.

As a result of considering the conceptual model it becomes clear that the quick-reference system is encompassed and given meaning by another system, aspects of which must also be considered if changes are to be made to the quick-reference system. The crucial aspects for the quick-reference system are how decisions about the path for an enquiry (quick-reference, extended reference or referral to subject librarians) are decided and what measures are used to monitor the results of an enquiry and the response of the user. The use of standards of some kind is implicit in the model which does not, therefore, represent a complete conception of how the system works. To show that another, encompassing, system is also involved a second boundary is drawn, which surrounds the original quick-reference system model.

There is a final step which needs to be taken with the conceptual model: this is to consider the overall process of Transformation from the point of view of its workability. Five related criteria, sometimes called 'the five Es', have been suggested for making this judgement:[5]

- efficacy – does the means of Transformation work?
- efficiency – is the use of resources optimized?
- effectiveness – is the Transformation meeting longer-term aims?
- ethicality – is the Transformation morally acceptable?
- elegance – is the Transformation pleasing from an aesthetic point of view?

For the model to be complete it is necessary to discuss how these criteria will be manifested and how suitable performance measures can be applied. The quick-reference system offers several problems which the senior management team would have to consider and which might, in themselves, promote thought. The evaluation of libraries and information services is highly problematic but some simplifying assumptions might be made. For example, it might be reasonable to assume that quick-reference questions would not be asked by students and faculty if their experience suggested that there was little chance of receiving an accurate and timely response; patterns of use might, therefore, be one measure of efficacy. Similarly, the range and type of questions asked might give some clue to efficiency: if most of the quick-reference questions are of the type which can readily be answered from a dictionary or encyclopedia, using senior staff to deal with them might not be an efficient use of resources.

The criterion of 'effectiveness' is the most problematic to assess within the milieu of libraries, information services and resource centres; simple reliance upon the notion of meeting 'longer-term aims' gives little guidance to managers. If 'efficiency' is concerned with carrying out a task to a particular level with the least use of resources, this criterion can be considered to be an internal measure relating to satisfactory arrangements for working; 'effectiveness' can then be considered as being concerned with the external relationships of the system: how well does it satisfy the needs of stakeholders? Using the 'stakeholder' concept does not contradict the notion of meeting longer-term aims because an aim for every service is to meet user needs within the context that has been decided, defined, or accepted for it by senior management.

Using the concept of stakeholder needs also encourages recognition that not all aspects of an organization are necessarily related directly to the needs of the users of the service. Although the 'user approach' is helpful in keeping in mind why services are provided, and makes explicit the ultimate location of accountability for their supply, it is sometimes difficult to link such an approach with the internal environment of a system. For example, judging the effectiveness of staff allocation to service points is certainly linked to the satisfaction of user needs but it has a much stronger link to satisfying the motivational needs of the staff concerned. Although dissatisfied staff will, ultimately, affect the quality of the service provided to users this will often be much more difficult to ascertain because of other factors which may also be involved. Seeing the staff concerned as stakeholders upon whom some of the responsibility for service quality devolves is likely to suggest to the manager more immediate ways of assessing effectiveness.

In working with a group which is considering effectiveness, therefore, the analyst should encourage a pragmatic approach which concentrates on stakeholder groups which are closely related to the problem or opportunity and then suggest the need to step back and consider how stakeholders with a less direct link are also affected. In this way, the concept of longer-term aims is clearer in definition and is also consistent with the user approach.

As far as the criterion of 'ethicality' is concerned, it is often difficult to structure discussion because what constitutes ethical behaviour may be unclear or contested. The analyst may prompt discussion by reference to professional guidelines or to a consideration of the beneficial nature of the proposal by asking, 'who benefits and is it fair to all who are involved?'

If 'ethicality' is difficult to decide, the criterion of 'elegance' is one which depends on highly individualistic judgements. It is, however, an interesting concept to discuss in relation to a project because that discussion will be rich in feelings and opinions as the participants share their individual Worldviews. Many of the assumptions built into the thinking of the participants about the process can become clearer and, by being open to debate, a much better understanding of the reactions of those who will use the process may become evident. From discussions about 'elegance' comes the development of a distinctive 'house style' and a concern with the quality of the process.

It is now time to return from the 'systems world' to the 'real world'; we entered the former with the idea that the quick-reference process might be one in which paraprofessional staff could be involved. Agreeing the root definition gave us a particular view which appeared to be a useful perspective from which to consider the problem further. Developing the conceptual model allowed us to identify the processes in the system and also highlighted links with another, encompassing, system which has an important controlling effect on the quick-reference system.

Stage 5: Comparison of models with reality

In returning to the real world, the insights gained from the systems world are carried with us to be tested against reality. This is a vital step because, although the systems world encourages a careful establishment of parts, processes, rules and controls, these may not necessarily be easily defined, recognizable or subject to change.

A major question in relation to each element in the conceptual model is whether such activity exists and, if so, in what tangible form? It may seem odd to be asking this question, as the models which have been developed have been based on discussion and analysis of real circumstances: how, then, can activities in them not exist? The need for the question arises because of a difference between a logical model, developed in the systems world, and reality. In the systems world the aim is to reveal what it is that starts a process, the decision points inherent in it, the rules used to reach a decision and the various outcomes, together with some idea of the probability of each. In the real world, decisions may be almost instinctive or subject to experience rather than the application of rules.

There is an example of the difference in the quick-reference case study. When a user asks a question the first action of the professional staff would appear to be to decide which type of reference service is required: quick-reference, extended reference or referral to the appropriate subject librarian. The logical, or systems, model implies that there are rules for arriving at a conclusion, but closer consideration reveals that these so-called rules are actually judgements based upon prior experience and knowledge of the Reference Library facilities. 'Quick-reference' is defined as a query which will take less than ten minutes research: clearly it is impossible to be sure that a particular enquiry will not exceed this time or that an 'extended query' will always take more than ten minutes. The decision rules are 'fuzzy' and could not be written down or taught in such a way that every member of the professional library staff would interpret them in the same way.

Other examples also abound in the quick-reference case study: judgement is at the core of the evaluation of the quality of the answer and of the user's response to it. This is also something which cannot be codified because it depends on the knowledge and sensitivity of the librarian and the willingness of the user to provide an evaluative response which is a true reflection of feeling.

Recognizing the difference between the systems model and the real world is especially important when change is anticipated. If rules cannot be codified then the need for experience and judgement will remain as central requirements for being able to carry out the process under consideration. Such qualities cannot be taught: they must be acquired and this suggests the need either for mentoring or for a suitably experienced person to make a decision prior to allocating the task to another person, perhaps a trainee.

Another point of comparison between the systems world and the real world is to determine whether there are significant aspects of the real world which have been omitted from the systems world model. What frequently happens is that the debate in the systems world concentrates on the central features of a process but procedures are forgotten when exceptional or special circumstances are encountered. Slipping back into the 'real world' allows the analyst and group members to examine or 'walk through' the model as if in real life and make sure that such distinctive aspects are remembered and recorded. This, in turn, may mean that parts of the systems world model have to be adjusted until a satisfactory fit between the two 'worlds' is achieved. This process of iteration is one of the hallmarks of good analysis and a competent analyst working with a responsive and committed group will find it easy to move from one world to the other, without disruption, as need dictates. As mentioned previously, going into the systems world need not be an esoteric discipline.

Stage 6: Determination of desirable changes

The purpose of soft systems analysis is to understand a problem or opportunity better in order to achieve change which is both attainable and beneficial. The process of analysis concentrates on both the Transformation and on the Worldviews of those involved but does not go deeply into the mechanics of the technology which might be used to improve the Transformation process. It also omits any formal considerations of cost. Both the technical and the economic aspects of a proposed change are, properly, to be explored using hard systems techniques, but it would be foolish to pursue the chimera of a change which is beyond the technical capacity or financial means of the organization. Therefore, part of the encounter with the real world must be a quick assessment of a change from the point of view of economic and technical feasibility. It is important to recognize that this is a preliminary screening and further work will probably be necessary to assess accurately the costs and technical implications. The main thrust of this stage of soft systems analysis is to identify changes which the culture of the organization will support and which will yield the desired improvement to the process of Transformation.

It is unlikely that only one way of improvement will emerge and the group needs to decide how to make a choice amongst two or perhaps several possibilities. Once again it is necessary to remember that soft systems analysis is concerned with promoting a better understanding of the human aspects of the organization: therefore the criteria for making a choice should reflect the interests of all the stakeholders involved. The root definition and rich pictures should be reviewed in order to ensure that each group or individual stakeholder is considered, and some weighting should be given to each. The weight chosen for each group or individual will depend on the perceived degree of involvement of the persons concerned with the outcome of the Transformation. Those closer and more directly involved should receive a greater weighting.

In the case of the quick-reference service, for example, any changes in the Transformation will affect the professionally qualified reference staff most strongly and, depending on the nature of the change, both users and perhaps those involved with other services of the Reference section may also be affected. For example, if it were decided that the more experienced paraprofessional staff should be allowed to answer some of the quick-reference queries this would have a strong and sustained effect upon both them and the professionally qualified staff. If the scheme is successful the users should not be aware of any difference in the quality of the response they receive to requests for help, but they might notice a difference in the attitude of the staff and, perhaps, an easing of congestion at the Reference desk.

In planning the change, it would be important to remember that the objective of the quick-reference service is to answer the enquiries of users and that maintaining or improving the level of satisfaction experienced by users must be the main concern when judging whether the change is successful. Once again, the value of the root definition is evident in reminding those initiating the change that there are groups other than the Actors which are affected by the Transformation process.

Stage 7: Choice of change strategy

Having chosen an action which will lead to a desirable change, the final step in the analysis can be taken: deciding how to implement the action. In the course of previous steps of the analysis, especially that of the choice of action, the method of implementation may already have become clear. There may already be experience or precedents which can be used and a feeling that the process of implementation is very well understood. Whilst such confidence may be helpful it may also lead to oversights and mistaken assumptions which could wreck the implementation and threaten the success of the change. For this reason the analyst should guide the group to consider the change strategy.

In the quick-reference case study, having decided to allow some paraprofessional staff to deal with queries, it is evident that the design of the new process must include suitable facilities for training and supervision of the designated paraprofessional staff, together with some means of assessing user satisfaction. The change strategy will specify who will be responsible for training, how and when, together with the resources required. The group may agree that there are three possible training routes: a course run internally by the professional staff of the Reference section, a course run by an external trainer but using resources in the library, or an external course to which the selected paraprofessionals would be sent. Factors such as the availability of finance and experience of training would be important in making a choice of strategy and the group should decide suitable criteria. If the aim is to minimize the external cost this may rule out the possibility of sending trainees on an external course; if it is to minimize the commitment of time by senior staff this may make the internally-run course unattractive. Other aspects to consider would be the mechanism by which queries will be screened and the quick-reference queries routed to an available paraprofessional and how, and at what point, the satisfaction of users with the progress of individual enquiries and the overall process should be judged.

In deciding a strategy, the group must consider several aspects of the process of change. The resources required and the people to be involved are obvious points for discussion. It is also important that the speed of implementation and its timing should be agreed. A means of monitoring the change is essential in order that effective control can be maintained; this, in turn, suggests the need to establish 'milestones', the passing of which should be the cause for a report or review of progress.

'Milestones' are of two kinds. One type represents a significant event, such as the completion of a particular part of a project; the other represents the passage of time spent on the project. By establishing milestones of both kinds, the use of resources on the project, its duration and links with other work in the organization can be kept under scrutiny.

If the project is large, involving many people, parts of the organization or an extensive change to a process, it will be helpful to subject the strategy itself to two stages of soft systems analysis: the building of a root definition and the resulting conceptual model. The focus of the Transformation will be on the means of achieving the required changes and the Actors will be those who are responsible for the design of the new process and putting it into place. The Clients will be those who will be working with the new process and the Owner will be the management team which is to be held accountable for the change. The Worldview will be that the process of change should be capable of being achieved whilst upholding the control factors, or milestones, established for the change. The Environment will consist of the resources available for the Transformation and whatever constraints may have been recognized.

By developing a root definition and conceptual model there is less likelihood of significant factors being omitted or of the provision of inadequate resources. Once again, use of the soft systems analysis approach should encourage discussion and the revealing of information; the task of the analyst is to guide and stimulate the discussion by providing a suitable framework.

Linking to hard systems analysis

Soft systems analysis is concerned with the human activity system: rich pictures, development of root definitions and conceptual modelling are means towards achieving a better understanding of the beliefs, values and behaviour patterns of the people involved with a process. Soft systems analysis is not directly concerned with whatever technology may be involved with the process, though the capabilities and limitations of that technology will be factors which form part of the Environment recognized in the analysis. The foundation of soft systems methods is the conviction that what is achieved in an organization is achieved by, with and through the people who are involved. Whatever changes to technology may be required it is essential, firstly, to understand the human component and how it might be affected.

The final stage of soft systems analysis is concerned with the choice of change strategy, as we have discovered. It is at this point that the link with other types of analysis, notably hard techniques, becomes most palpable. Having defined the nature of the change required and the effects that the change may have on people, this understanding now needs to be combined with whatever technological considerations have become apparent. A complex project for which completion on time is a vital consideration may suggest the need for scheduling using a hard technique such as Critical Path Analysis (CPA), for example. If the project involves the implementation of new computer systems then the development of a full system specification will be the next stage. The decision to introduce a new benefits package leads to the need to organize job evaluation studies.

In this way, the process of analysis moves from the soft to the hard milieu: the move represents a change of tools, the choice being guided by what has been learned by using the soft approach.

'Soft' and 'hard' are opposites in meaning but this should not be taken to imply that the modes of analysis to which the adjectives are attached are divorced from each other or wholly unlike and incompatible. The need to move into the 'hard' milieu is signalled by the recognition that the focus of the analysis is moving, for the time being, from the Human Activity system and its attributes to the study of the technology associated with it. This move need not be confined to the end of the stages of soft systems analysis: the move can be made whenever the group feels confident that it has a clear and correct understanding of the human elements. It is also possible to move temporarily into the hard mode to gain a better understanding of a technological aspect and then return to the soft mode for further consideration of the implications for the Human Activity system. It is also worth noting that the move to the hard mode is not always necessary as a complement to the analysis already performed: some problems or opportunities are solely concerned with the behaviour taking place in the Human Activity system.

Further episodes of analysis

It has earlier been noted that the model of soft systems analysis is not cyclic. This should not however be taken to mean that only one episode of soft systems analysis is required for the exploration of every problem or opportunity. One of the advantages of using the soft systems approach is that it is possible to use as much, or as little, of it as seems necessary for better understanding. Similarly, it is often beneficial to apply the analysis in several episodes in order to gain clearer insight into the project forming the subject of the analysis.

Checkland and Scholes have drawn attention to the frequent need for successive episodes in their discussion of the 'stream of cultural enquiry'.[6] The first episode, 'Analysis One', is concerned with defining three crucial roles: that of 'client', 'owner' and 'solver'. Being clear about the identities of the people or groups occupying these roles is vital for the conduct of the study and for the implementation of whatever change is decided upon. Lack of clarity will quickly lead to confusion and avoidance of action, because accountability has not been assigned and agreed. A rich picture is a good way of capturing the interests and responsibilities which the people who seem to be involved express and is also useful for promoting discussion about areas for which responsibility seems unclear or contested. In some cases, a person or group may legitimately assume more than one role: for example, 'owner' and 'solver' are often the same, especially if the study is of a process the effect of which is limited to one part of an organization. It is important to ensure that the reasons for attributing these roles to individuals or groups are discussed and clearly under-stood: in particular the distinction between 'actor' and 'owner' needs careful debate. Because the owner has caused the study to take place it may be supposed that the wishes and expectations of the owner should be used to define the scope of the study and the direction of change; the analyst must recognize, however, that the study is being carried out on behalf of the owner in order to solve a problem or take an opportunity which is in the domain of the 'actor'. An 'owner' will tend to have considerable authority over the use of the resources of the organization, but the 'actor' will have delegated responsibility for those which are most immediately concerned with the subject of the study.

In the quick-reference case study, for example, the 'owner' is the Reference Librarian, who is responsible for the direction and supervi-sion of staff and activities, while the 'actors' would be the staff of the Reference Library. Whilst the view of the Reference Librarian would be an important consideration in the discussion of what changes could be made, the staff should have a more informed and detailed view of the work involved in quick-reference and will be ultimately responsible for ensuring the selected changes take place smoothly and effectively.

In addition, the Reference Librarian is an 'actor' in relation to the 'owner' of the Reference Library system, who is the University Librarian. Responsibility for resource use rests with the University

Librarian, who must be accountable to the university authorities. Which level of 'owner' is recognized as being significant in a study will depend on the nature and extent of the solution developed. If it can be handled within the resource framework of the Reference Library, the owner is the Reference Librarian; if it has an impact beyond this point, it falls within the province of the University Librarian.

As can be seen, the role of owner is one which needs careful consideration and it is often helpful to regard initial definitions as being tentative, revisiting them periodically to make sure that they are still appropriate. Just as the focus of a study may change with greater insight, so too the attribution of roles may have to be reconsidered. Before implementing change, however, those roles should have been clearly defined and mutually accepted.

'Analysis Two' is concerned with the behavioural characteristics of the people involved with the Human Activity system under study. In Chapter 3 the characteristics of groups were briefly explored. 'Analysis Two' encourages the analyst to explore the social characteristics in terms of roles, values and norms of behaviour. Because these three characteristics have a powerful influence on group and individual behaviour an appreciation of these characteristics will facilitate understanding of how best to achieve change and evaluate the responses of the individuals involved. For example, in the quick-reference case study, if it is evident that the professional staff lay a strong emphasis on 'professionalism' as a value, it will also be possible to anticipate their resistance to paraprofessionals being involved in answering queries. The analyst may have to prompt a discussion of what 'professionalism' might mean and ask if it is only a consequence of possessing professional qualifications.

Roles, values and norms change throughout the life of an organization: the intervention of the analyst and the process of analysis itself will lead to their modification. Thus the information and perceptions gained by the analyst must be regarded as provisional. The analyst will also, almost certainly, have to gather the information indirectly: asking for a description of the norms or values of a group may be met with incomprehension or a partial statement. Further contact will enable a sensitive analyst to detect far more through behaviour and comment and to correct or expand earlier perceptions.

The 'political' dimension of an organization constitutes the focus for 'Analysis Three'. In any Human Activity system there will be occasions where progress has to be made through arbitration and the accommodation of differing interests. We have seen that a key factor in the working of groups is the recognition and exercise of power; an analyst exploring a problem or opportunity with a group must be aware of the location of power in the group, how it is expressed and how it is being used.

Organizational politics is also bound up with less visible matters, such as acceptance of membership in social groups which do not appear on any organogram. The need to understand how people relate through such groups is important in explaining how the organization works and recognizing how change has to be negotiated. In one organization with which I was involved, following a reorganization the members of one department were absorbed into another. These members referred to themselves as 'the Old Gang' and often expressed a collective view quite at odds with that of the rest of the department; successful change often required a separate discussion with this group. Only after most of the members had left the organization and been replaced by new staff not sharing this collective memory did any sense of unity begin to appear.

The task of handling and using the information gained through Analyses One, Two and Three is delicate. It is, primarily, information to be used by the analyst in deciding an appropriate strategy for increasing the group's understanding of the problem or opportunity being studied. It may be helpful to feed back some of the insights into the behaviour of people in the organization, but this needs careful preparation if the participants are not to view this as being in some sense punitive. The characteristics being considered are of the deepest significance to the identities of people and their relationships with each other. It should not be surprising that discussion will inevitably promote the expression of strong feelings.

Analyses One, Two and Three are therefore conducted as an adjunct to the analysis of the project under consideration and are regularly reconsidered as the analyst gains fresh insight into the working of the organization.

Conclusion

A complete episode of soft systems analysis has been followed through in terms of both theory and example. Not every exploration using soft systems analysis will follow the same pattern nor necessarily use the same sequence of steps. All, however, should reveal a similar underlying design: the trigger for an investigation is a sense of unease or the expectation that something could be improved. The investigation moves on to a better definition of what is needed and an understanding of the range of human behaviour, values and attitudes which surround the definition. Joint consideration of the definition and the human component leads to the selection of a path to change and a move, if needed, to the consideration of technological matters using the techniques of hard systems analysis.

The process of analysis is never, in an absolute sense, complete, because no situation remains static. From time to time it is beneficial to return to analysis of the Human Activity system, especially if there have been radical alterations such as a change of personnel. However, for practical purposes there must come a time when action displaces analysis: the best expression of this is a plan for change to which the group responsible for implementation feels a sense of commitment. Commitment is an expression both of intellectual satisfaction and emotional security.

References

1   Ackoff, R. L., 'The art and science of mess management', *Interfaces*, **11**, 1981, 20–6.
2   Checkland, P. B., *Systems thinking, systems practice*, Chichester, Wiley, 1981, 154–5.

3   Mushkat, M., 'Problem definition in social planning', *Journal of applied systems analysis*, **13**, 1986, 97–108.

4   Checkland, P. B., *Systems thinking, systems practice*, Chichester, Wiley, 1981, 317.

5   Checkland, P. B. and Scholes, J., *Soft systems methodology in action*, Chichester, Wiley, 1990, 288.

6   *Ibid.*, 44–51.

Chapter 5

The tools of soft systems analysis (I): interviews

THE MAIN TOOLS of soft systems analysis are the interview and model building. Their use is so frequent and central to the analytic approach that it is worth considering them in some detail.

The craft, or skill, of systems analysis is bound up with knowing how best to use the tools and not simply with being able to recognize what they are. In addition, 'craft' implies the need for practice rather than solely the digestion of theory. With either of the tools to be discussed in this and the next chapter, experience and reflection on use will lead to confidence and an improvement in technique. Feedback from participants in model-building and interview sessions will reinforce self-analysis if the analyst is receptive and open to comments.

Tools need maintenance and a critical, reflective approach to their use: it is easy to fall into bad habits or to adopt patterns of behaviour which provide an obstacle for other participants. The analyst is peculiarly vulnerable, for example, to the temptation to becoming an over-active participant during discussions, seeking to fill gaps in conversation or to lead the thinking of others. It is salutary occasionally to allow an observer to review the behaviour of the analyst during interviews, discussions and other events and report back on the form and structure of the interchanges which took place.

Tools also need regular use, both to keep them in good condition and to allow users to refine their skills. The skilled analyst has a variety of tools which can be selected because they are suitable for particular jobs but most will also develop core expertise in using a small set and will tend to rely upon them. The danger is of becoming so familiar with a particular tool that it tends to be used irrespective of whether another might be more appropriate in certain circumstances.

It is a good discipline occasionally to pause, review the circumstances, and consider which tools of analysis are especially suitable.

Interviewing: a moral tale

In Chapter 3 the conditions associated with working with a group were thoroughly explored. Such conditions form the basis for an understanding of how and when to interview participants.

Interviewing is always a group activity: even if the analyst is engaged in a one-to-one interview the duo represent a small group, called together on a temporary basis, and will respond to each other in terms of a group dynamic. In other words, the interpersonal behaviour represented by either party will have an overwhelming influence on what is said and what is understood. Elsewhere it has been explained that the analyst cannot assume a detached, Olympian position of simply asking questions in a manner which is value free. The attitudes, disposition and demeanour of the analyst will have just as much effect on the results as will those of the person being interviewed.

To illustrate this I should like to discuss an incident which arose early in my consulting career. It exemplifies some of the problems of interviewing and also illustrates how the analyst can perturb a stable system.

The request had come from the director of a large library service with several branches. A first-time automation project was in progress: computers had been installed in the central service-point and many functions had been successfully changed from manual to computer-supported form. Attention had now switched to the branch service-points and it was here that the first glimmering of a problem with the project began to emerge: many of the branch personnel appeared to be resistant to the idea of automation even though the supposed benefits had been explained to them. The Director was of the opinion that they were scared of the notion of using computers; nowadays the word 'technostress' might be used to describe this condition. I was asked to conduct a short 'computer familiarization' training programme, centring this on one of the larger branch service-points.

In order better to understand what the issues might be, and discreetly to check that the perception of the Director was correct, I asked to spend two days going around the branch service-points. This gave me an opportunity to meet all the staff and to discuss their understanding of the automation project.

The circumstances of each visit were almost identical: I had asked to meet as many members of staff as could be spared from their duties and we then tried to discuss the objectives of the project and how it would affect their service-point and work. I was hoping to discover that many were unaware of how computers could be used in libraries and would benefit from a short training course. This was, however, not the case: the staff in every service-point had a good general knowledge of computer use, had seen examples of automated services in libraries elsewhere, and did not display alarm at the prospect of running similar services. They were, however, reluctant to discuss the rearrangements which would be necessary in their service-points in order to accommodate the computer peripherals and other units. It was difficult for me to understand why this was so but I did not wish to antagonize by asking a direct question about this seeming reluctance.

A clue to the problem came in one of the larger branches towards the end of my tour. The librarian in charge of the branch (who I had met before at a conference) was quite voluble in his criticism of the approach taken by the team responsible for planning the automation. In particular, he recounted an incident where the project manager (a member of the senior management team of the library service) had visited the branch accompanied by the senior member of the installation team and had then proceeded to tell the staff of the branch where the equipment was to be installed, what changes would be made to the counter area and when this would be done. 'It was as if he owned the place,' the librarian remarked. 'In no sense was there any consultation or attempt to take into account what we thought about the plan. No one even asked us for our ideas.' The other staff agreed and several commented that this was not an isolated incident:

in fact, the attitude of senior management of the library service was considered almost dictatorial and their relationship with the branch service-points was seen as remote. This view was consistent with a few remarks that had been made at other service-points on my tour and served to explain the apparent resistance to automation: because there was no real attempt at seeking their opinions, they recognized that they could have little effect on the process.

The problem was not solvable by the comparatively easy means of providing a training course: it concerned a lack of communication between senior management and those in branch service-points, sufficiently serious to have a sustained and deleterious effect on the motivation of branch staff. In order to assist the library service it was necessary to resolve this problem and promote improvements. The temptation simply to point out what appeared to be the real problem to the Director was strong, but I considered that this would have little effect because its tenor rang counter to the impression given me by the Director in an earlier interview: the Director had remarked that the staff of the library service were a unified team. 'Just one big happy ship', was the Director's way of describing them. The interviews had revealed that this was certainly an illusion but it was also one on which the Director placed some considerable emphasis; it was clear that cohesion was a strong value for the Director and an outright suggestion that it was not a characteristic manifested by the staff could lead to rebuttal and denial.

Instead, I suggested that there was a need for a 'training' day but that the content should not be confined to explaining the functions of computers: it should include an opportunity to discuss different designs and arrangements for the counter and fittings. 'But hasn't that all been done?' asked the Director. 'Yes, but the branch staff feel that they would have benefited from the opportunity to compare designs used at other branches,' I replied. The Director agreed that the designs for each branch had not been finalized and that no serious delay would arise from allowing this additional exercise to take place. I suggested that it would be helpful if he and the project manager could attend and this was accepted.

The 'training' day started with my giving a brief overview of the way in which computers could affect the work of staff and of how, in turn, users might be affected by computer use. Because these events were taking place at an early period in the development of computer applications in libraries in the United Kingdom, what was described was relatively unsophisticated but it was apparent that staff and users would find considerable change in their patterns of work. The staff, in particular, would find a difference in the pace of work as the manual issue system was replaced by the use of bar-code scanning. I concentrated on this and asked the staff present at the training day to consider how they would, ideally, want their work space arranged in order to make the new system as comfortable to use as possible. In an adjoining room the project manager had constructed a mock-up of a typical circulation control desk with the new equipment in place; this enabled staff to handle the unfamiliar bar-code scanners and try the relative merits of using the equipment when sitting and standing. By good fortune a member of staff at the central library, which had already been using the new equipment for a few weeks, was present and able to offer some comments on the good and bad points of design of the workspace installed there.

After a mid-morning break I asked the staff from each branch to take an outline plan of the public area of their library and to consider the placement of the circulation desk and its approximate layout. The project manager and the Director moved amongst the groups and offered advice and comments. By the end of the morning the staff of each branch had worked out their needs and I suggested that, after lunch, each branch should comment briefly on its design and invite comments and suggestions. This session proved very useful because the characteristics of a general design began to emerge as the staff debated the merits of different approaches. At the end of the session, there was clear agreement on a design which could be modified to accommodate the architectural vagaries of some of the older buildings. Most importantly, there was a much improved feeling amongst the staff and some evidence of enthusiasm had replaced the earlier sense of sufferance.

At the final session I suggested that we discuss this experience of design. As might be expected, the participants were quite enthusiastic but no one mentioned the apparent transformation in their attitude or commented on the communication problems which had seemed to be at the base of their previous feelings. It was the Director who finally raised the topic by asking whether the participants believed that a similar approach to problem solving should be used on other occasions. This prompted a discussion of the perceptions the branch staff held of senior management and the prevailing impression that senior management tended to be remote and rather out of touch. The Director acknowledged that this view was a serious criticism of management style and the project manager accepted that there had been little involvement of the branch staff in the development of an implementation plan for the computerized system. It was agreed that it would be premature to plan improvements in communication at the present meeting, but that it would be preferable to arrange another workshop with communication as its specific focus. That workshop, I later learned, was very successful. Once participants had accepted that communication had to be two-way and that failures could be attributed both to the senior management team and to the branch staff there was a concerted move to lay down guidelines for better practice in this respect.

Interviewing: a critique

The purpose of the foregoing case study is not to present the analyst as hero: this could not be the case because, on reflection, there were several errors in my approach. It serves to illustrate, however, the way in which the role of analyst can be used to shift the orientation of participants from confrontation to examination of the reason for their differences. As discussed in previous chapters, this is the true function of analysis.

In retrospect, my errors in working with the individuals and the group were considerable and, given less forgiving circumstances, could have had serious consequences. They also had, as we shall see, a moral dimension.

The errors concern my willing acceptance of there being a communication problem and of this being the 'real issue' rather than there being a training need. In doing this I was placing greater value on my own perception than on the greater experience and knowledge inherent in the Director and senior management team. Moreover, I was persuaded to give the voluble branch librarian greater credence because he was an acquaintance. Thus there was the possibility of a double mistake: of wrongly rejecting the establishment view and of wrongly preferring the view of someone I knew. As it transpired, it was probably correct to reject the view that training was the only factor which needed attention because the evidence gained in prior interviews tended to support this conclusion. However, in accepting 'communication problems' as being the true concern I was treating this view as if it were a revelation, complete and hardly needing further checking. Instead of retracing my path to check this assumption with some of the previous interviewees, I had gone on with it firmly in mind and unhesitatingly behaved as if it were proven, rather than being a provisional explanation requiring further study. That it received apparent support from the Director in our later discussions did nothing to mitigate the potential danger of this error; because both of us had been puzzled by the listless character of the responses from the branch staff it was a relief suddenly to be confronted with a model which seemed to explain the reason and which offered an objective problem which might be solved.

The point should be emphasized: all models, apparent truths, theories and explanations are only provisional statements, the continuing credibility of which depends upon their being perceived as consistent with other information, coherent and testable. The temptation is to forego the latter task of checking when the former conditions seem to be satisfied.

The moral dimension concerns what, in theological discussion, is sometimes called the 'Principle of Reserve': the notion that the communication of doctrine is best achieved in small steps so that a solid bedrock of belief is established before difficult, and more profound, doctrines are explained. In setting up the training session I had not informed any of the participants, including the Director, that the ostensible objective of the session (computer familiarization and discussing some aspects of workplace design) cloaked the quite different purpose of exploring improvements in communication. My prior contact with the staff of the branches had concentrated on discussing their needs for a course on computer applications in libraries but the pattern of the training day quickly left this topic behind. The design exercise created the link between reviewing applications and considering communications by forcing participants to exchange and discuss ideas with senior management. In the course of this, all participants rediscovered their willingness and ability to listen to each other. The question is: was this oblique approach to solving a problem correct?

It is pointless to justify one's actions by maintaining that it all came right in the end. In not revealing the true nature of the 'training' I was placing greater faith in my ability to resolve conflict than in the abilities of the participants to solve problems once the causes were better understood. I had abdicated my proper role as analyst and usurped that of problem owner. This was an untenable position because I was neither employed by that library service nor contracted to provide consultancy services beyond their immediate training needs. Because I was an 'absentee owner' there was no way in which I could discharge the responsibilities I had unwittingly assumed and it was fortunate for the participants that the intervention of the Director caused 'ownership' to move seamlessly back to its rightful place.

Ground rules for interviewing

From the foregoing discussion it must be apparent that one of the foundations of good interviewing is to be truthful both to oneself and to those one is interviewing. This means being open about the purpose of an interview and informed about the background, as far as is possible. Clearly this also requires the interviewer to be well prepared and this constitutes the first and most basic rule. Aspects of the preparation will take different forms, depending on whether the interview is with one individual or a group and whether it is a first encounter or one of a sequence with the same interviewee or group.

There are three dimensions of interviews which provide the basic framework for preparation. I shall call these 'logical context', 'logical depth' and 'logical width'. Each needs separate consideration in planning interviews but final consideration needs also to be given to the set as a whole. This is entirely consonant with the general approach of soft systems analysis, which encourages a unified rather than a disintegrative approach.

Logical context of interviewing

'Logical context', as a dimension, considers the relationship between the interviewees and the problem or opportunity which forms the centre of the study. The concept of the stakeholder has already been introduced in an earlier chapter: when thinking about a particular activity or sequence of events it may be possible to distinguish degrees of involvement amongst stakeholders and stakeholder groups which could provide a clue to the depth of feeling each may be willing to express. Thus, in thinking about the work of a group of cataloguers, the 'logical context' may be a department which is the workplace for the group and the problem or opportunity may be that of moving its location. The stakeholder groups include that of the cataloguers but also include groups in neighbouring departments who may have an interest in the space to be made available or occupied: planners of the building, construction people, maintenance and removal workers, the departmental head and immediate line manager and the financial controller. All will have some interest in what is being proposed but that interest will be engaged at different times and by different aspects of the planned change. Those with an immediate and continuing interest are the cataloguers, whose working conditions will be affected by – and during – the move, and who have a continuing interest in their future surroundings. The locus of the 'logical context' concerns first the present working arrangements and space and then moves to the future working arrangements and space. The interests of the other stakeholders and stakeholder groups will represent different aspects of the move, such as cost control, planning and scheduling of alterations, impact on neighbouring departmental space and staff but the locus remains the same.

A sense of the logical context is important when preparing for an interview because it assists in identifying potential stakeholders and their likely range of interests but, at the same time, this needs to be balanced against the danger of assuming too much. The description of a problem or opportunity will give some clues as to the locus of the context but this must be regarded as a provisional identification which will be tested during the interviews and may have to be modified.

As an example of the need to remain open to the possibility of a change of locus, consider again the problem of the departmental move. Initial interviews might indicate that the members of the department have no qualms about the move and that its impact on neighbouring departments is well understood and not regarded as problematic. The move is regarded as an uncontentious episode by the stakeholders who have been interviewed. The locus of the logical context remains as originally identified until an interview with the person responsible for supervising the necessary building work reveals that there will be a considerable problem in meeting the proposed timetable. At this point the logical context moves to the department responsible for building work, in order to identify the extent of the problem and its estimated impact on the timetable of the project. The locus may remain there if further study suggests that there could be ways of rescheduling its work and alleviating the timetabling problem. The principal stakeholders have become the members of the building work department and the logical context is also now in that department. The other stakeholders and stakeholder groups, such as the staff of the cataloguing department, are still involved but have moved to a position tangential to the logical context. If further study reveals that no adjustment of scheduling is possible then the locus returns to the cataloguing department in order to consider the implications of a delay in completing the relocation.

Consideration of the locus of the logical context is important if the analyst and the participants are not to experience a feeling of overload and helplessness arising from the developing nature of the problem or opportunity. If inclusiveness and holism are virtues of soft systems analysis, then one of its potentially problematic effects may be the sudden expansion of a problem or opportunity into a 'mess' which it is difficult for anyone to feel confident about managing. Recognizing the locus, even if it eventually shifts as the nature of the problem or opportunity changes, enables the analyst to distinguish between those stakeholders or stakeholder groups which are central to a solution and those which are, for the moment, on the periphery. This does not mean that their input can be ignored or not sought, but that factors which are critical to a solution are, for the most part, resident elsewhere. Consideration of a complex problem can be structured by the analyst without losing sight of the whole.

The effect on interview planning and technique is subtle. In order to use soft systems methods an analyst must be prepared to spend time in gathering the distinctive perspectives of stakeholders but this does not necessarily mean that each stakeholder or stakeholder group will have an input which is central to the logical context. The time afforded each stakeholder and the depth of discussion about the central issues will vary according to the degree of involvement. For those who are located within the logical context the exploration may be expected to be more searching, the interviewing more probing and the analysis more structured. For those outside the logical context the level of discussion will be less probing. If the locus shifts, then the centre of attention will also shift and those previously on the periphery will be brought into a more searching debate.

The physical domain is only one way of typifying the logical context. There are many issues for which a consideration of physical location is unimportant or irrelevant. A proposal to introduce a new stock selection policy necessarily has more to do with an emotional and professional perspective than with the work of a particular department, although its adoption may eventually affect the work of acquisitions and will also have an impact on users of the library. The locus of the logical context is with those who have strong involvements with the existing policy and will have immediate dealings with the effects of the change.

It is also possible to recognize accountability as forming a logical context although, on its own, it is not a prime determinant of who is centrally involved in a problem or opportunity. Similarly, possession of resources may be used as a logical context though it would be unusual for this to be the only reason for locating the logical context with an individual or group; the focus of interest is usually on how the resources are to be used rather than the resources themselves.

In summary, proposing the locus of the logical context of a problem or opportunity allows the analyst to consider who should be included in the preliminary sessions and the direction and issues which the interviews might take. As a result of the preliminary interviews it is quite possible that the initial identification of the locus will be revealed as incorrect, unhelpful or misleading; until the preliminary work has been done the locus should be regarded as provisional. As mentioned earlier, it is also possible for the locus to shift as consideration of the problem or opportunity matures and fresh insights or additional factors alter its nature.

Logical depth of interviewing

Consideration of the logical context leads on to an exploration of the second dimension, that of 'logical depth'. By this is meant the inherent complexity of the issues surrounding a problem or opportunity.

'Complexity' is a measure of the depth of the rule base governing a particular problem or opportunity. Here, the notion of 'rule' is used in a general sense to cover formal rules (such as might be found in conditions of employment), laws (both natural and as social constructs), procedures which have been sanctioned by an organization or group of individuals, and norms of behaviour which provide the social framework within which groups and individuals respond to each other. The rules need not necessarily be written but they should be explicit in the sense that their existence can be recognized, they can be identified by those involved in the interviews, and their effects can be adjudged.

To describe a problem or opportunity as 'complex' is to suggest that there are many levels of rules and that the outcome of an event at one level will result in a chain of events at other levels, the outcomes of which may not necessarily be immediately obvious or predictable. A complex problem has logical depth if, and only if, the rules can be ascertained and their operation understood in sufficient detail for predictions to be made about outcomes, with a high degree of success. It is quite possible for a problem to be complex but to lack logical depth because the rules are unknown, unknowable or the outcomes appear arbitrary. The operation of the weather system, for example, is highly complex, but because few of the rules are known and remain, in the present state of knowledge, unknowable, it lacks logical depth.

A problem with great logical depth might be that posed by the need to develop and introduce a procedure for evaluating the collection of a library. The rule base is complex because there would need to be agreement over what aspects of the collection should be evaluated, what measures should be used, how the data should be collected, what the results might mean and in what way they would be used to adjust policy. Each level of decision interlocks with the following level and a change at one level can have an effect on levels above and below. In addition, the effect of a change can be predicted: decide to include in-library use of material as one of the aspects to be evaluated and there are immediate questions for the deeper levels on what is practicable and likely to be useful. If it is decided to reduce additional workload on staff to the minimum, for instance by relying on data derived from an automated circulation-control system, there will be questions about how this will affect the idea of measuring in-house material usage.

Consideration of logical depth gives the analyst clues about the need for detailed discussions of procedures, the need to establish which rules are already recognized and explicit and which are implicit but need to be made apparent. It will also aid in identifying areas where rules will need to be constructed: in the foregoing example, the way in which results will be used to guide the collection development policy will probably need to be considered further.

A problem or opportunity with great logical depth will most likely require expert consideration once the framework and guiding principles for a solution have been established. The analyst will find that meetings of large groups are, almost certainly, going to be unable or unwilling to handle the necessary detail in order to establish a coherent set of rules. Smaller task teams and consultation with individuals will be necessary, with the purpose sharply focused on particular aspects of the venture. Part of the work of the analyst will be in keeping a close check on the results of each deliberation, looking for effects on the work of other groups and linkages to other levels of the problem or opportunity. The role of the analyst will change for a time from a facilitating to a coordinating activity; this will continue until the rule base has been explored, at which point the role of the analyst will change, once again, to facilitating the work of the groups in a general, perhaps plenary, session at which the complete rule base can be revealed.

A problem or opportunity without great logical depth may be intrinsically simple, in the sense that the rules are well known, unambiguous and without significant linkages. A decision to set aside funds for a special project, for example, has administrative implications such as the need to open an accounting entity for the administration of the funds, and it may have an impact on the distribution of funds to other budget heads. However, the rule base for dealing with this is likely to be well known and the decision to make the allocation will be uncontentious, unless it is disputed by other budget holders. Even if it is, the need for debate will revolve around principles, ideals and ethical matters rather than a need to understand the rules of the system. The analyst will find that this sort of problem or opportunity needs discussion at a plenary, inclusive, level in order that the feelings and agendas of each stakeholder can be recognized and addressed. Facilitation of this needs a plan which emphasizes interaction and sharing of ideas rather than the expert examination of rules.

Some problems or opportunities exhibit an apparent intransigence which derives from an unknowable rule base rather than its lack of depth. A discussion about the possible effect on behaviour of violence portrayed in the media is an example: the case for such a link is strongly contested but the evidence for either side is unclear. Add this to a consideration of the need for censorship as part of an acquisitions policy in a public library and the problem of developing that policy is, essentially, insoluble. The rule base for developing and applying an acquisitions policy is clear but a fundamental assumption of such a policy is that there will be agreed and clear criteria for making a choice of materials. The unknowable aspect of the rule base surrounds the ability to recognize and categorize 'violence' and the supposedly bad effect of including material portraying such images in stock.

Configuring interviews to deal with such topics is especially difficult. Because the rule base is unknowable there can be no clear agreement on what constitutes a solution, how it should be reached and how this condition will be recognized. The role of the analyst in planning will be largely concerned with establishing a forum for debate and ensuring that the participants first explore what is going to signal agreement. It is almost certain that participants who have thought out their own position on such issues will not relinquish those positions – unless they are susceptible to emotive arguments – but they may be willing to abide by a majority view whilst still reserving their own case. What the analyst has to prepare the group to accept is that there cannot be agreement but there may be acquiescence or provisional acceptance of a working arrangement.

Another type of interview planning is required for a second instance where the logical depth of a problem is not great. Such circumstances arise most commonly when dealing with a clash of personalities, interpersonal problems or the need to deal with long-running disputes. Whatever the antecedents may have been for such problems they are probably now unclear or forgotten. What group members are dealing with and, indeed, fighting about is not truth or reasoned argument, but the desire to dominate. 'Facts' are not an important element in the argument though apparent facts may be advanced during the course of events; the focus is on feelings and the pressure is on participants to declare or confirm their loyalties. The analyst must appreciate that during such disputes there is a lot of energy being released and there may well be benefit for the participants in having an opportunity to discharge their feelings. The process needs to be carefully managed if the end result is not to be destructive. In order to prepare for such meetings the analyst must meet each of the disputants separately in order to understand their points of view but not to offer support, encouragement or agreement. What the analyst is seeking are points over which there is no dispute and towards which the participants can gradually be led once their feelings have been expressed. It may be necessary for the analyst to carry out several rounds of individual consultations, by way of preparation, before convening a group forum. The group process then becomes one of discharge of feelings followed by the establishment of points of agreement and the rebuilding of a relationship around such points.

The dimension of logical depth presents the most challenging area for the analyst in planning interviews. The analyst is, after all, an agent whose task is to guide individuals and groups through topics which are difficult either because of their complex rules or because of the absence of those rules. Thorough preparation means being aware of the nature of the arguments and sensitive to the personal characteristics of those involved.

Logical width of interviewing

In the previous section the idea of complexity was considered as a characteristic of logical depth. Logical width in relation to interviewing also has an associated characteristic: it is the contrast between the attributes of simple and of compound interaction, when applied to the involvement of the groups of people associated with the problem or opportunity.

Simple interaction means that there is one person or group of people for whom the problem or opportunity is a central concern: involvement of other people or groups is confined to the implementation of the decisions of that central person or group. In other words, people or groups outside the central person or group will not become engaged in cryptic or contentious activities as a result of decisions made at the centre.

An example of simple interaction would be the need to reorganize the flow of work in a department. The central group consists of the people working in the department, including the supervisor; the problem to be explored is for the group to agree on a change of working practice. With the proviso that the output of the department is not changed in terms of either quality or quantity, no one outside the department is involved. If, however, quality or quantity are likely to change as a result of the revised procedure then other groups and departments might themselves be affected: the simple nature of the interaction then moves towards a compound involvement.

Another example might concern changes to the technology used by a department. The group centrally involved will consist of those who will use the new equipment; agreements should be made on the type of equipment needed, a programme for installation should be developed and the level of training that will be needed should be defined. Other people and groups will be involved but, if their involvement is to supply equipment or services, such as training, according to a plan previously agreed with the central group, their input is still simple. It is only if one of the supplier groups is unable to fulfil its obligations, or if there is some contention about form and manner of supply, that the problem becomes compound and thus expands the logical width.

Notice that there is nothing in the foregoing to suggest that the solution to a problem or opportunity being considered is, in any sense, inherently easy or uncontentious. That characteristic arises, as previously described, from a consideration of logical depth. In terms of logical width what is being considered is the spread of involvement in the intricacies of the problem or opportunity.

By contrast, the compound nature of involvement describes a problem or opportunity where there are several individuals or groups, all actively involved in assessing a decision, making it, or suffering its consequences as distinct from simply implementing it. The distinction is subtle, because often the people making a decision are also those who will be implementing it; it is, however, a contrast which needs to be made because it typifies a difference in the degree of personal involvement and identity with the problem or opportunity. Taking part in the implementation without some additional degree of involvement usually implies a low degree of interest in the process of the discussions prior to reaching a decision.

Problems or opportunities with a large logical width imply, therefore, circumstances where there are several individuals or groups each of which will need to be involved in the process of consultation and discussion prior to decision making. A highly compound problem or opportunity implies that there are many such individuals or groups and that they have interacting and intersecting areas of interest which may have a bearing on the problem or opportunity.

It is possible for the logical width to increase or decrease during the course of a round of discussion. Factionalism may splinter a hitherto unified group, thus increasing the compound nature of involvement; unification of different points of view may also result in a coalescing of groups and thus a reduction in logical width. In each of these cases there is the need to distinguish between a cosmetic change and a sustained change of identification with the discussion. A cosmetic change represents an often temporary move, more concerned with tactics than with any fundamental shift of priorities or viewpoint. A cosmetic change does not connote a change in logical width of any great significance and the analyst may well discover the position abandoned, or changed yet again, in the course of a discussion. A sustained change represents a fundamental re-orientation, sufficient to affect the essential dynamics of the discussion.

The case study again: a counter-critique

Putting the dimensions together provides an opportunity to look once more at the case study which opened the chapter, to discover if they can render any additional insights into what went wrong with the conduct of the interviews. At the time of the events described I had not developed the 'dimensions' model, but subsequent analysis of it persuaded me of the need for a framework to use when planning interviews.

The starting point of my investigation was the interview with the Director of the library service. In terms of logical context the problem seemed clear at the outset: it was located in the branches and concerned training needs. The logical depth also seemed slight because the training need was apparently well specified and did not extend beyond a simple introductory course to familiarize branch staff with the use of computers. The logical width could safely be regarded as narrow because one unitary group, albeit composed of staff from different branches, was implicated and no other group or person, other than myself as trainer, would be involved. The starting position appeared propitious, with a well located problem of no great complexity which involved one clearly defined group.

The value of the framework would have been in alerting me to some very odd characteristics of the problem when set against the dimensions. Considering the logical width, the apparent ability of the Director to typify the staff at the branches as a group distinct from the staff at the central service-point was interesting and should have provoked me to question the extent to which they had already been involved with the automation plans. The apparent isolation of the two groups, central and branches, should have seemed odd as an initial identification and prompted questions about communication. It is almost inconceivable that a large change such as automation should not have had an early impact throughout the whole library system, at least in terms of awareness of its progress. If there really was lack of awareness then that should have signalled communication problems as being characteristic of the library system.

The assumption that the logical context of the problem could be located in the branches should also, on reflection, have seemed odd because it suggested that, in relation to the apparent problem of training, it was possible to deal with the branch staff as a category distinct from those at the central service-point: but why, and in what way, could the branch staff be regarded as different from those at the central service-point? To what extent did the staff at one branch communicate with those at another?

Given these misunderstandings on two dimensions it is scarcely surprising to find that the third dimension, of logical depth, was also wrongly described. The need for training is usually a complex issue because an effective training programme should take account of the present knowledge of the participants, the degree of change they are expected to encompass and the effect which is to be achieved. Talking to a group about the use of computers would have achieved little, other than keeping them amused for a short time and providing some information. What, however, the Director had implied was the need for a considerable change in attitude on the part of the branch staff: this is not something which is susceptible to education alone. The problem, even in its initial formulation, was poorly understood by both the Director and myself.

Considering the problem afresh, and applying the framework of the dimensions, it is possible to characterize it in the following way. The initial formulation should have suggested that the logical context was located principally with the personnel section of management and concerned the assessment and description of a possible need for training coupled with concerns about communication between staff at branches and the central administration. The logical width of the problem then depends on the degree to which the groups of staff at the central service-point, those at each branch and the senior management, have a unitary view of the benefits of automation and are able to share such a view. If there are already the beginnings of a commonly held view, then the logical width is narrow and discussion is probably all that is required to resolve any problems; if there are significant differences in points of view, misunderstandings and misperceptions, then the logical width is greater and each distinct group will need preliminary work to identify the areas of difference and the reasons for them before discussion can begin.

Apart from the changes in the locus of the logical context and the uncertainties about the logical width, it is the logical depth of the problem which changed dramatically during the course of the case study. The initial identification of a training need should have prompted questions about how the need had been identified, but further consideration of this was overtaken by the revelation about lack of communication and the implied resentment felt by some of the staff in the branches. The logical depth of the problem immediately increased because what was actually needed was a complete shift in the pattern of communication and consultation, both between each branch and the central administration, and between branches. This inevitably makes the problem more complex because it implies changes in protocol and decision making; such changes may have an impact on the time taken to complete projects and on the skills and commitment required of senior staff. It will also have an impact on the way in which senior staff perceive their roles: a shift from direction to facilitation could be required.

In train with the change in logical depth is a large change in logical width. The unitary view, with its simple acceptance of a single group, must be discarded. The staff of each branch and the central servicepoint represent, at the very least, several separate groups. There may well be more than one group at each branch, each having a distinct view of the need for improved communication; senior management represents at least another.

Considering the initial problem statement within the dimensional framework would have alerted me to the likelihood of a mistaken identification having been made. This should have prompted me to a much more searching discussion with the Director; it should also have suggested the need for specific questions when visiting the branches about the quality of communication and their feelings about relationships with senior management at the headquarters.

So far the critique of the case study has suggested that knowledge of the dimensional framework would have enhanced my intervention and reduced the possibility of a blunder being made. A reasonable counter-critique is to suggest that, with the benefit of greater experience and insight, I could have avoided such a blunder in any case and there is no need for the framework. Consultancy and analysis are crafts, however, and practitioners have to develop and use tools which suit them individually. This does not prevent them from learning from each other and from the collective experience of the profession. For me, the dimensions provide a vital first step in considering the nature of the intervention required to assist with solving a problem, or in developing a strategy to make use of an opportunity.

A final point about the dimensions must be made. As a technique of intervention the use of soft systems analysis relies upon the analyst directing activities without leading people to solutions. The dimensions assist in defining what kind of thinking space is needed for the investigation, who should inhabit that space and what issues need to be addressed.

Conducting interviews

Discussion is the main way in which analysts gather information and make that information available to the other stakeholders concerned with solving a problem or making best use of an opportunity. The dimensional framework outlined above facilitates preparation for such interviews but it is also important to consider the actual process of interviewing.

At first sight there is nothing intrinsically difficult about talking to people: we do it all the time. However, this statement underlines the main problem: the main task in soft systems analysis is the much more difficult one of *listening*. Encouraging others to talk and to address the issues thought relevant to the discussion will require some initiation by the analyst but should not form the dominant part of the proceedings.

Interviews, whether individual or group, depend for their success on the participants being at their ease. The starting point is for each interviewee to be made aware of the reason for the interview, its location and expected duration, well before it is to take place. It is helpful to summarize this in a briefing document but the analyst must be careful to make this informative but not leading, especially in respect of suggesting solutions or even the boundaries of the problem or opportunity. It is the views of the interviewees in respect to these matters which are being sought: supplying too much information in a briefing document may lead to a narrowing of focus and the implementation of defence mechanisms in advance of the interview.

The location of interviews also needs careful consideration. Interviewing within the soft systems analysis culture is somewhat different from the practice to be found in hard systems thinking: within that milieu the focus of interest is to discover details of work practices, technology used and other administrative details. For this purpose it is often helpful to interview people within their workplace so that they can demonstrate activities or refer to documentation.[1] The soft systems technique is concerned with understanding the problem and is therefore not so tightly focused on specific job details. Releasing the thinking of people from their habitual bounds is often aided by choosing a location away from the normal place of work. At the least it is important that interviewees be free of the distractions of telephones, electronic mail and colleagues and the demands of their everyday work. Neutral ground, such as a meeting room, can be used if it is not possible to use a location away from the normal place of work. Choosing neutral ground also ensures that all participants feel equal and can share in the experience of finding a group identity and working together.

Any interview, whether in the hard or soft systems milieu, is an exercise in interpersonal behaviour. The participants may know some or all of the other members of the group and the analyst may similarly be known or new to them. In either case, the analyst has to be on the look-out for signals which suggest boredom, distress, tension or non-engagement with the proceedings. The discussion of group behaviour in earlier chapters suggests a variety of behaviours which groups may exhibit as a means of avoiding or trying to renegotiate awkward aspects of a discussion. Individuals may also use similar ploys. Whilst it is not for the analyst to force participation it is important that other approaches be tried if it becomes apparent that further progress using a particular approach is being impeded. Note needs to be taken of the obstacle, however, because it must be approached again, perhaps from another direction, so that the reason for the obstruction can be better understood.

In order to facilitate the discussion the analyst can create a 'script' of questions and statements thought to be apposite, from the initial consideration of the problem or opportunity against the framework of logical dimensions. The script can only ever be a starting point and should not be used to control or curtail discussion: it is intended to ensure that the participants explore the problem or opportunity as fully as the available information allows. Subsequent meetings will build upon this information and the better understanding gained through earlier discussion. The preparation for each meeting should include the construction by the analyst of a fresh script taking account of the results and insights gained from previous meetings.

Considering the emotional and power balance in a group is also important. It is rarely possible to anticipate how the members of a group will settle and work together unless the analyst knows the participants from previous work; adjustments may have to be made as experience of the response of each participant is gained. The strongest guide which can be used during the initial selection of groups is the recognition that mixing people from greatly different levels in an organizational culture which lays emphasis on hierarchy is rarely productive. In other organizations where the culture is more collegial, and based on recognition of expertise rather than position, this may not be such a problem, but even then some care is needed if those with a more junior status or with less recognized sources of information and experience are not to feel intimidated or overawed.

Mixing participants from groups who are known to be in the midst of a disagreement will lay the foundations for an emotional meeting. Although such a meeting may eventually be needed it should be planned as such and should certainly not be combined with an initial exploration of a project. In such circumstances it is best to keep participants in separate groups until the nature of the disagreements is more fully understood.

Using the dimensions in planning the conduct of interviews

The locus of logical context provides a good guide to who should be invited to participate in the interviews. Initial discussions may indicate that the locus has been misplaced or will change as more information about the project is discovered. Thus the selection of participants must be regarded as provisional. The locus indicates the main area from which questions concerning the project will arise and this helps to define the stakeholders: each stakeholder or stakeholder group should be considered as a potential participant because their input will have a bearing on the shared understanding of the problem or opportunity involved. If this is not the case then they have a peripheral rather than a central interest and their interests are not located within the logical context.

Logical width, whether narrow or wide, serves to identify the range of stakeholder groups and whether they are connected in a simple or compound manner. If the connection is considered simple, then meetings as a single group should be productive, so long as interpersonal characteristics do not prove disruptive. For dealing with a project where involvement is considered compound there will be a need for meetings of each identifiable group before a plenary meeting will be helpful. The purpose of the separate meetings is for the analyst to gain a better understanding of the similarities and differences of approach. If there is evidence of nascent agreement then the next step can be to bring the groups together into a plenary meeting at which the analyst can identify the points of agreement as a basis and then persuade the participants to address the areas of disagreement. If, however, there is evidence of disagreement about fundamental aspects of the project it is advisable for the groups to remain apart until the analyst has gained a clearer understanding of the reasons for the differences. If these arise from misapprehensions then the analyst can seek to adjust these in a future meeting of the group; it is best that the groups do not meet in a plenary until the base for a satisfactory shared understanding has been established.

Preparation for interviews is more difficult if it becomes evident that there are apparently irreconcilable views between the groups. In the first instance the analyst will be the only person completely aware of the extent and nature of such differences, although in the ordinary course of events in any organization or social grouping, rumour and informal sharing of information may have spread knowledge of the difference in views. In the second, the analyst is faced with a considerable temptation to use this overall knowledge to lead the separate groups towards a solution by minimizing the differences. Such manipulation will almost certainly be counterproductive because no firm foundation of agreement has been established. What the analyst should be preparing to do if the differences are irreconcilable is to bring the groups together in the shared knowledge that there *are* major differences. At this meeting there should be discussion of whether the project should be abandoned, the opportunity rejected or the problem acknowledged as presently insoluble, or whether the plenary members would be content with taking a decision through a mechanism such as a majority vote.

Logical depth can only be fully appreciated as discussions and information become more freely available. The attribution of degree of logical depth to a problem or opportunity is, in the first instance, provisional. Because it is a reflection of the rules and interactions implicit in the system under consideration, it also indicates the extent to which rules and details will be essential to the solution. The analyst will have to prepare for interviews concerning projects with great logical depth and complexity by being fully aware of the rule base revealed so far and the loose ends which still need to be tied into that base. The use of flowcharts and decision tables drawn from the hard systems milieu[2] can be the best way of recording and working with such information. For projects of shallow logical depth the implication is that it is not so much agreement on rules as shared knowledge of feelings and perceptions which will be important in creating a satisfactory solution. The drawing of rich pictures is often the best way to work with such information. The analyst prepares for this by considering what questions will draw out such feelings and beliefs, the strength with which they are held and the extent to which they are shared by others in the group.

The axes of dimensions are usually considered to be orthogonal and, by implication, independent of each other. However, within the context of soft systems analysis I have used them as an aid to thinking about problems rather than as a mathematical abstraction. It should not be surprising, therefore, to find that alteration on one dimension also often, but not necessarily, influences the placement on another. Thus a change in logical depth to something of greater complexity often increases the logical width, because a deeper rule base tends to include impacts on other groups. A change in logical width to reflect a more highly compound involvement often brings with it a shift of the locus of logical context: as more stakeholder groups are involved so the context tends to spill over to encompass several areas.

Types of interview

The final consideration in planning interviews is the type of interview to be used. There are two choices to be made: structured or unstructured and individual or group. Once the logical dimensions of the problem or opportunity have been considered, the type of interview needed may be readily apparent; if not, it will be necessary to make an explicit decision.

Structured interviews require a script which is closely followed. The analyst uses the script to elicit information, often of a factual nature, to confirm conclusions and to prompt consideration of a range of notions which the analyst has decided, provisionally, to be of significance. The script is worked out in advance by the analyst and consists, essentially, of a questionnaire with a few open-ended questions to promote discussion. The purpose behind this is to guide the investigation quite strongly through areas where the information is readily available and not likely to be contested but where the use of questionnaires completed on their own by participants might inhibit the revelation of additional information relevant to the discussion. The approach is fairly economical of the time of participants and is particularly suited to the initial stages of an investigation where facts and the rule base have to be uncovered. It is of limited value in assessing perceptions or feelings however, and should not be used when dealing with information or statements known to be contentious.

Unstructured interviews do not lack a script, but it is less apparent and consists of questions which are open-ended and deliberately encourage participants to talk. Preparation for this type of interview is quite demanding because the analyst has to consider the various directions in which the interview may go as a result of the information elicited, and be ready with salient questions to prompt the participants. The skill of this type of interviewing is to be able to provide space and encouragement for discussion whilst at the same time ensuring that the range of topics which needs to be addressed is covered. A quick wit and experience will add to the versatility of an analyst under such conditions but the need for careful preparation is not superseded.

Individual interviews are often used as the starting point of a study, especially with senior management, in order to gain insight into facts and the probable shape of the problem or opportunity. The focus is on individual characteristics and interpretation, coupled with a need to understand the power relationships in the organization. Individual interviews may also be used later in a project as the logical width expands to incorporate other stakeholders or stakeholder groups. If it is possible to identify leaders, whether legitimate or arising from one of the other power bases identified in Chapter 3, an interview with them will often be sufficient to define the input of the group.

Group interviews are chosen when the essence of the information required comes from the quality of the interaction between participants. This may be a means of assessing reactions, exploring reasons for disagreement or of building solutions; the key point is that more can be gained from the group behaviour than from the responses of the participants as individuals. In preparing for such interviews, the analyst must anticipate and think through the consequences of disagreement, as well as agreement, and work out how best to facilitate discussion. The starting point is invariably a statement of intent from the analyst followed by a series of questions arising from general reflections on the project under discussion or the findings of previous meetings.

Conclusion

Interviewing is a critical skill for analysts; it is a tool which needs constant practice to master. Its fascination lies in the essential novelty of every occasion on which it is used. There is, for the reflective practitioner, always something to learn.

References

1   Underwood, P. G., *Managing change in libraries and information services: a systems approach*, London, Bingley, 1990, Chapter 4.
2   Daniels, A. and Yeates, D., *Basic systems analysis*, 3rd edn, London, Pitman, 1988, 30–52.

Chapter 6

The tools of soft systems analysis (II): models and model building

O N FIRST BEING introduced to the use of soft systems analysis the response of many people is to remark that it seems little different from participative decision making because so much emphasis is placed on involvement of the participants. There is considerable truth in this statement provided it is acknowledged that the focus of soft systems analysis is how to achieve better understanding of a problem or opportunity rather than on making decisions.

It is the activity of model building which typifies soft systems analysis, at least in the minds of many participants; its importance has already been referred to in Chapter 2. At the end of an interviewing session the tangible evidence of what has gone on is often represented in a model which has been built up during the course of the session. The model serves as a record but it is much more than this: it has also been instrumental in stimulating thinking, clarifying, assisting participants in discussing feelings and perceptions and in sharing those views with others. For the analyst it serves the double purpose of being a statement of present understanding and of giving clues to avenues which await further exploration.

All models are necessarily a simplification, but this is also a virtue because the reality of life is so often a 'mess' of conflicting aspirations, impressions and ideas. Models allow the participants to identify the form or essence of what they are discussing in order better to understand how the components fit together and the rules by which they interact.

Model building is such a frequent part of managerial work that it is surprising that its use in soft systems analysis often attracts comment. The reason may be that the models which are most often encountered by managers tend to be schematic, as in the case of a budget statement or the sequence of subjects in a classification scheme, or plans of buildings and equipment. These are formal representations of ideas, produced according to rules and requiring some skill to interpret. The models built during soft systems analysis appear not to be of this kind in that they are free-form, often include cartoon-like pictures, and do not appear to conform to a set of rules. Whilst many people 'doodle' as an aid to thought, such pictures rarely form part of their records of events or meetings: yet this is exactly how the models developed during a period of soft systems are intended to be used.

Whilst many people find the use of the models helpful this is certainly not true of everyone and, since the models are an aid to thought rather than an essential part of work within the soft systems milieu, there need be no insistence upon their use if the analyst or the participants feel uncomfortable with them. Equally, the form of a model and the technique of its drawing are individual and can be modified to suit the circumstances, skill and requirements of the participants.

What must be recognized, however, is that even if it is decided not to draw a representation, a model will emerge from the discussions of participants and will need to be documented in some way. The activity of model building is part of the mental preparation of the participants and analyst in order better to understand the Human Activity system which is under investigation.

What is a model?

Simply defined, a model is anything which represents those parts of reality considered essential for some purpose. It stands in the place of reality in order that we can think, negotiate, play or experiment. In using a model in this way we do not seek to suspend our appreciation of reality, but to gain a better understanding of how the characteristics we have chosen to include in the model perform together. A model, then, is not a substitute for the reality but an abstraction observer. Migration needs to be managed, however: it is a responsibility of the analyst to be aware of potential change in status and to foster debate about the identity of the stakeholders and the nature of their involvement. Failure to take this responsibility seriously can have several potentially deleterious effects. People who should rightly have been acknowledged as having stakeholder status may have been omitted on purpose or through oversight: those so excluded may then cast doubt on the legitimacy of any decisions reached. The continued inclusion of people who no longer have a stakeholder interest may allow them to adopt a 'maverick' role, uncontrollable and not accountable to the rest of the group for their opinions and behaviour; it may, on the other hand, transform into boredom, a condition which is equally destructive of group processes. The role of disinterested member is, in fact, something of a misnomer: to be within a boundary one should have an interest capable of being defined by reference to the indicators.

Reviewing the position of the boundary and why it has been drawn in a particular position may be of occasional value in revitalizing a group. Commitment to a project can wane if the analysis and planning extends over a long time, but can be revived as participants discuss why and to what extent they are stakeholders.

Cognitive maps

Processes or products can be represented using cognitive maps. There are a few rules to their construction and consequently they can take many forms. All share one characteristic, however: the abandonment of the linear form of development characteristic of most note-taking.

A move towards using mapping as a means of recording, or depicting, discussions feels risky because so many of the words of the discussion seem to be omitted from the resulting diagrams. Linear note-taking, which often resembles the minutes of a meeting, imposes its own discipline however, which may conflict with the creative responses of a good group discussion. 'Hang on, just let me finish writing that down', or 'We mentioned that earlier; let me find the note' – these are examples where the process of recording has become intrusive and serves to control the pace and direction of discussion.

The use of cognitive mapping can be much less of an obstacle to the free flow of discussion. In addition, a well-conceived map can provide the analyst with many reminders of the discussion and facilitate the writing of full notes and minutes if these are thought necessary.

Buzan[2] has discussed a number of styles which are suitable for diverse purposes such as keeping records of discussions, taking notes whilst reading, and preparing lectures. I have used the technique for many years and have found it to be of considerable value for all of these purposes and also of help in facilitating a group discussion.

The technique depends on quickly deciding what is the central topic under discussion and describing this at the centre of a piece of paper, overhead projector transparency or writing board, using one or two words. This word, or words, is then enclosed in a 'bubble' and, as a new aspect of the topic emerges in the discussion, this is noted using a word or two inscribed on a line emerging from the bubble. Comments arising from this aspect are linked to the line with other lines, each such line also carrying a one or two-word summary. If another aspect is introduced it is recorded on a new line emerging from the bubble.

Figure 6.1 illustrates the record that might emerge when documenting a discussion. The discussion group consists of the members of the work group in the stock ordering department of a library. The analyst prompts discussion when necessary and keeps a note of progress using cognitive mapping on a writing board. The map is visible to the participants and is used to summarize points emerging in the discussion. The record commences with a definition of the central purpose of the discussion: to investigate the working of the stock ordering department of a library. This is noted at the centre of the bubble with the one word, 'ordering'. Several topics arise from this, covering the staffing complement of the department, the origin of requests, their frequency of reception, the problems which arise and

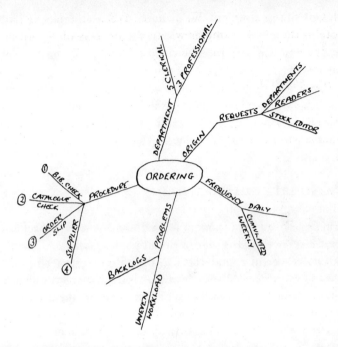

Fig. 6.1   *Cognitive map of the 'stock ordering' department*

the procedure adopted to deal with the requests. No order is implicit in the diagram except at one point: the 'procedure' line ends in a 'fan' showing a sequence of tasks. This is documented by numbering the arms of the fan to show in what order each task is accomplished.

The technique, sometimes called 'structured note-taking', is very flexible and allows the discussion to move and develop without constraint or the imposition of a linear sequence. If, whilst discussing procedures, a participant refers back to a problem (for example, inaccuracies in the bibliographic information on requests) this topic can simply be added to the 'problems' line without upsetting the pace or direction of discussion.

Another purpose of such maps is to stimulate discussion and tease out points needing clarification. For example, 'uneven workload' and 'backlogs' are noted as problems; the 'frequency' line shows that the practice in the department is to cumulate the requests received each day and process them in a weekly batch. The analyst might encourage the group to consider whether some of the problems might be attributed to this 'batching' procedure and to explore the consequences of moving to a system of daily processing. This discussion would constitute a new central topic and could be documented by starting a new map, with 'daily processing' as the central bubble.

The Buzan method can be learnt quickly and adapted to many circumstances. It is a basic and powerful technique.

Personal construct maps

Cognitive mapping techniques based upon other approaches have also been developed. That described by Eden, Jones and Sims[3] provides a powerful complement to Buzan's general technique and allows a more detailed analysis of the Worldview and resultant theories developed from experience of that world by the participants. I shall describe the basics of a form of 'personal construct mapping' which they use; readers interested in the elaboration of this form are referred to their book.

The essence of this personal construct approach depends on considering a problem and identifying the attendant ideas, beliefs, values and attitudes, together with the relationships which are perceived between them. Taken together these constitute a personal or group 'theory' of the world in relation to the problem.

To begin such a personal construct map, the participants have to decide on a name for the problem. The discussion of a suitable name is a crucial first step which may be surprisingly difficult. With a discussion involving several participants the Worldviews held by each are almost certain to differ and so the concept of what the problem is will also vary. The initial identification and naming of the problem must be tentative, therefore, and as the map emerges in the course of the discussion there is every chance that the perception of the problem and its name will change. This does not represent a failure of the mapping: rather, it is the function of the map to facilitate discussion and refinement of views.

which will be of value for a certain length of time and a limited purpose.

Two main classes of models, characterized by the purpose for which each is intended, can be identified:[1]

- process models
- product models.

Process models describe the way in which a decision can be reached or a task achieved. They make explicit reference to the relationships to be found amongst those stakeholders and groups who will be making the decision or will be responsible for task achievement, thus focusing on what might aid or impede progress to these goals. A timetable for a project is a familiar example of a process model.

Product models are used to describe the 'Worldview' of the participants, identifying the components, how they may be grouped and the relationships between the groupings. An organogram of the team responsible for a project provides an example of a product model.

The example of a project has been used to illustrate the difference between the two classes of models. It will also serve to exemplify another point: the classes are not necessarily mutually exclusive. The timetable for a project cannot be developed without some firm idea of the composition and skills of the project team; similarly, the parameters set for the timetable may well influence the size and organizational pattern of the project team.

The usefulness and quality of a model are critically dependent upon what features of reality we have chosen to include. A model of an aeroplane can, for example, be made of a single sheet of paper: it will have wings and a body; given a certain amount of skill in its making and launching, it will fly. For the purpose of demonstrating the phenomenon of aerodynamic lift it is adequate. It would, however, be quite unsuitable for the purposes of an aircraft designer or a collector of model planes, both of whom might be much more interested in the accurate reproduction of features of the fuselage than in a primitive ability to demonstrate flight.

Whilst one may accept this idea of simplification when discussing physical models, the process of abstraction may not be so evident when it comes to dealing with iconic or schematic models; even so, the process of simplification is an active part of the process of building such models. Consider the floor plan of a building: reality is suddenly represented by only two scaled dimensions and many of the physical features evident in the 'real' surroundings, such as the damp patch on the floor by the window, may be omitted. In presenting a set of financial accounts, the individual flows of income and expenditure are summarized and represented by overall amounts. The reality of the actual amounts, their timing and impact on cash flow week by week, is concealed.

Just as the stakeholders involved in a discussion have different views of a topic, so the decisions they may individually make about what features of reality need to be represented in a model will also vary from individual to individual and over time. To continue with the example of financial accounts, the financial manager may decide that a more stringent approach to cash planning is required, and he will focus on the timing of income and expenditure: the analysis of cash flow becomes the essence of the model and the representation of reality will change accordingly. This does not mean that the first model is now incorrect, inaccurate or of no further use: it is simply that it represents another view which is not appropriate for the problem in hand. Model building must always reflect purpose.

It is also perfectly possible for there to be two, or several, models in use at the same time. For example, the floor plan of a building is a two-dimensional model which can be complemented by a series of isometric drawings giving a three-dimensional view. These, in turn, might be augmented by a physical scale model in which the layout of each floor can be seen in relation to the others, which together constitute the building, and the vertical connections between floors such as lift shafts, ducts and stairwells can also be seen. Each model, or representation of reality, has a particular value and can offer insights not easy to gather from the others; conversely, each can offer only a limited view of reality. The key points are that, provided the models are accurate in their representation, none of them is in any degree better or less approximate than another, and that each will represent the world from a particular point of view.

Models also introduce another important aspect of soft systems analysis: indeed, an issue which is of importance in any kind of work with systems. This is the process of establishing a boundary, to be considered next.

Sorting out boundaries

Consider again the model plane used to demonstrate aerodynamic lift. It is a simple model to make, economical in its use of resources and, if made well, a surprisingly good flyer. This success is derived in the main from decisions made almost instinctively, about the boundaries of the model. A successful paper aeroplane tends to be almost all wing, starting from the nose and flaring out steadily to the rear. The body of the plane has been reduced in the model to a link between the wings and a convenient point to hold when launching the plane. The model is mostly about the wings: the boundary of the model is defined by them. Experimenting with wing shapes or changing the size of the model will not alter this boundary but may provide some insight into the relationship between size, wing shape and aerodynamic lift. The model remains centred around wings.

Now add a paperclip to the body of the plane. Addition of a small amount of weight alters the behaviour of the model in flight and by experimenting with the position of the paperclip some significant improvements in performance can be made. As well as adding physical materials to the model, the use of the paperclip has changed the boundaries of the model. The model now enables the tyro aeronautical engineer to experiment with body weight, centre of gravity and wing shape.

Boundaries are defined by components, functions and the relationships between them. They serve to group together what is considered relevant to a particular phase of an investigation and to screen out what is not perceived as relevant. In the course of an investigation of a problem or opportunity it is of the utmost importance that the analyst and other participants have a clear idea of the boundaries of what is being studied; in addition, the analyst must recognize that such boundaries are always provisional and subject to change as better understanding is gained or as investigation reveals the need to include additional components. Part of the work of the analyst is to be aware of the dynamics of the boundaries and to draw the attention of the group to significant changes.

The concept of a boundary has been introduced briefly in Chapter 4 and mentioned in Chapter 5. Just as the dimensions are a fundamental consideration when planning interviews, so the recognition and mapping of boundaries is an essential part of model building. Boundaries in physical models are clear and tangible and many other boundaries are familiar because they are conventional: the functional divisions apparent in the organization of libraries, information services and resource centres are a good example of a set of boundaries which are accepted and almost unconsciously used in discussing work.

Some boundaries are more subtle in meaning, however. A social group recognizes, or even creates, the means of distinguishing between member and non-member, using recognition signals, indications of status and shared behaviour. Besides being quite intangible, such boundaries may be difficult to perceive but are fundamental to a clear understanding of the behaviour of the individuals, both members and non-members.

The analyst should always be wary of accepting the placing and position of a boundary without some justification. A boundary divides and excludes in ways which may once have made sense in the life of an organization but may no longer represent anything other than convention. Organizational memory is long and practices and procedures once adopted are sometimes retained without the necessary critical attention: they become a matter of convention rather than reflecting a continuing need. An interesting example of this is the concept of the 'library without walls', a development which reflects the impact of information technology. The recognition that in a networked environment, in which many information sources can be held in electronic form on a file server with access being effected over the network, the concept of a library as a central repository for information must be rethought, is typical of the fundamental reorientation which becomes possible once a boundary is removed. At the

observer. Migration needs to be managed, however: it is a responsibility of the analyst to be aware of potential change in status and to
foster debate about the identity of the stakeholders and the nature of
their involvement. Failure to take this responsibility seriously can
have several potentially deleterious effects. People who should rightly
have been acknowledged as having stakeholder status may have been
omitted on purpose or through oversight: those so excluded may
then cast doubt on the legitimacy of any decisions reached. The continued inclusion of people who no longer have a stakeholder interest
may allow them to adopt a 'maverick' role, uncontrollable and not
accountable to the rest of the group for their opinions and behaviour;
it may, on the other hand, transform into boredom, a condition
which is equally destructive of group processes. The role of disinterested member is, in fact, something of a misnomer: to be within a
boundary one should have an interest capable of being defined by
reference to the indicators.

Reviewing the position of the boundary and why it has been
drawn in a particular position may be of occasional value in revitalizing a group. Commitment to a project can wane if the analysis and
planning extends over a long time, but can be revived as participants
discuss why and to what extent they are stakeholders.

Cognitive maps

Processes or products can be represented using cognitive maps.
There are a few rules to their construction and consequently they can
take many forms. All share one characteristic, however: the abandonment of the linear form of development characteristic of most
note-taking.

A move towards using mapping as a means of recording, or depicting, discussions feels risky because so many of the words of the discussion seem to be omitted from the resulting diagrams. Linear
note-taking, which often resembles the minutes of a meeting,
imposes its own discipline however, which may conflict with the creative responses of a good group discussion. 'Hang on, just let me fin-

ish writing that down', or 'We mentioned that earlier; let me find the note' – these are examples where the process of recording has become intrusive and serves to control the pace and direction of discussion.

The use of cognitive mapping can be much less of an obstacle to the free flow of discussion. In addition, a well-conceived map can provide the analyst with many reminders of the discussion and facilitate the writing of full notes and minutes if these are thought necessary.

Buzan[2] has discussed a number of styles which are suitable for diverse purposes such as keeping records of discussions, taking notes whilst reading, and preparing lectures. I have used the technique for many years and have found it to be of considerable value for all of these purposes and also of help in facilitating a group discussion.

The technique depends on quickly deciding what is the central topic under discussion and describing this at the centre of a piece of paper, overhead projector transparency or writing board, using one or two words. This word, or words, is then enclosed in a 'bubble' and, as a new aspect of the topic emerges in the discussion, this is noted using a word or two inscribed on a line emerging from the bubble. Comments arising from this aspect are linked to the line with other lines, each such line also carrying a one or two-word summary. If another aspect is introduced it is recorded on a new line emerging from the bubble.

Figure 6.1 illustrates the record that might emerge when documenting a discussion. The discussion group consists of the members of the work group in the stock ordering department of a library. The analyst prompts discussion when necessary and keeps a note of progress using cognitive mapping on a writing board. The map is visible to the participants and is used to summarize points emerging in the discussion. The record commences with a definition of the central purpose of the discussion: to investigate the working of the stock ordering department of a library. This is noted at the centre of the bubble with the one word, 'ordering'. Several topics arise from this, covering the staffing complement of the department, the origin of requests, their frequency of reception, the problems which arise and

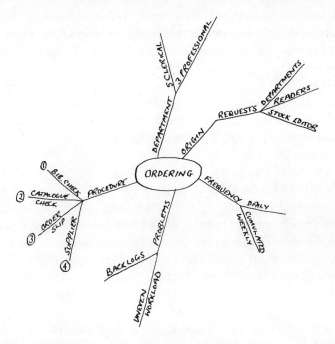

Fig. 6.1   *Cognitive map of the 'stock ordering' department*

the procedure adopted to deal with the requests. No order is implicit in the diagram except at one point: the 'procedure' line ends in a 'fan' showing a sequence of tasks. This is documented by numbering the arms of the fan to show in what order each task is accomplished.

The technique, sometimes called 'structured note-taking', is very flexible and allows the discussion to move and develop without constraint or the imposition of a linear sequence. If, whilst discussing procedures, a participant refers back to a problem (for example, inaccuracies in the bibliographic information on requests) this topic can simply be added to the 'problems' line without upsetting the pace or direction of discussion.

Another purpose of such maps is to stimulate discussion and tease out points needing clarification. For example, 'uneven workload' and 'backlogs' are noted as problems; the 'frequency' line shows that the practice in the department is to cumulate the requests received each day and process them in a weekly batch. The analyst might encourage the group to consider whether some of the problems might be attributed to this 'batching' procedure and to explore the consequences of moving to a system of daily processing. This discussion would constitute a new central topic and could be documented by starting a new map, with 'daily processing' as the central bubble.

The Buzan method can be learnt quickly and adapted to many circumstances. It is a basic and powerful technique.

Personal construct maps

Cognitive mapping techniques based upon other approaches have also been developed. That described by Eden, Jones and Sims[3] provides a powerful complement to Buzan's general technique and allows a more detailed analysis of the Worldview and resultant theories developed from experience of that world by the participants. I shall describe the basics of a form of 'personal construct mapping' which they use; readers interested in the elaboration of this form are referred to their book.

The essence of this personal construct approach depends on considering a problem and identifying the attendant ideas, beliefs, values and attitudes, together with the relationships which are perceived between them. Taken together these constitute a personal or group 'theory' of the world in relation to the problem.

To begin such a personal construct map, the participants have to decide on a name for the problem. The discussion of a suitable name is a crucial first step which may be surprisingly difficult. With a discussion involving several participants the Worldviews held by each are almost certain to differ and so the concept of what the problem is will also vary. The initial identification and naming of the problem must be tentative, therefore, and as the map emerges in the course of the discussion there is every chance that the perception of the problem and its name will change. This does not represent a failure of the mapping: rather, it is the function of the map to facilitate discussion and refinement of views.

As an example of how this can work, consider the case of the stock ordering department previously described in Figure 6.1. 'Uneven workload' was a problem which was identified. Suppose the group decide that this is a problem which requires more detailed study. The analyst puts this problem name at the centre of a writing board and invites discussion about what it means. What is uneven workload and how is it recognized? It may soon become apparent that two quite different problems could be described. One is an unevenness in work*flow* characterized by sudden rushes of work followed by slack periods. The other is an inequitable work*load* characterized by some people having more to do than others in the department. Perhaps the group decides that it is the latter which is the better description of the problem as they perceive it.

Once a name for the problem has been established, the participants should be encouraged to think of a state which they believe would have the effect of removing the problem as named. At first participants may respond by suggesting the 'received truth', the apparently favoured means of resolving the problem, rather than indicating what they really think would be a solution.

To illustrate this, consider the response to the notion of 'inequitable workload': the 'official' answer may be that the allocation of duties by the supervisor of the department should be improved in order to even out the distribution of tasks and this could be shown on the map as 'improved allocation by supervisor'. Further probing by the analyst may reveal that there is a belief that some members of the department are lazy and do not put sufficient effort into what they are doing. The choice of the word 'inequitable' offers the clue: it carries feelings of unfairness and grievance that the term 'uneven workload' does not.

A proposal to include on the map words carrying an emotional overtone should always be used to prompt discussion. The essential point is to allow participants to discover, and to acknowledge, their concern for the matter and why they have such a concern. To continue the example above, the reason for the concern may not be a feeling that everyone should have exactly the same amount of work to do: instead it may be prompted by a feeling that the supervisor is favouring some people by giving them a lighter allocation or substituting more interesting work in their allocation in place of some tedious tasks. Although this explanation indicates what may appear wrong with the allocation system it still does not indicate the reason why the group members are concerned. The question, 'Why does this matter to *you*?' needs an answer.

Further discussion, prompted and moderated by a critique by the analyst, may be necessary before an explanation is forthcoming, especially if the group, or individual members, have an emotional investment in what is being discussed, or fear reprisals for stating their views. The analyst should listen carefully to *what* is being said and *how* it is expressed, then look for expressions of agreement and disagreement, or a silence which can indicate either assent or the withdrawal and distancing of an individual from the views of others. Sometimes the emotional discomfort is so great that the analyst may feel it is best to conclude the group discussions at that point and to continue them during interviews with individuals or smaller groups. Whatever strategy is taken it is important not to lose sight of the point at which the emotional temperature rose: this is usually associated with the nub of the problem.

In the example, (Figure 6.2), the directions taken by the discussion are indicated by two paths emerging from the initial statement of the problem and its supposed remedy. The left-hand path relates to the feelings of the work group and contrasts perceptions of laziness with the need for improved motivation. The consequence of improved motivation should be a better working relationship whilst the consequence of this not developing may be a high staff turnover. The right-hand path refers to the role of the supervisor. An improvement in allocation of tasks would solve the problem of inequitable workloads but is inhibited by the supervisor having favourites. The remedy for this is for the supervisor to allocate tasks according to abilities of the work group members; the consequence of this will be a better working relationship. By linking the statement groups with arrows the analyst can show how the group perceives each statement in relation to another and can also indicate how one set of circumstances may cause another: 'better working relationships' arise as conse-

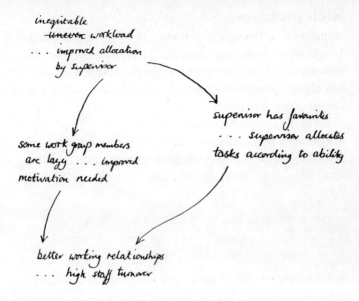

inequitable
~~uneven~~ workload
... improved allocation
by supervisor

Some work group members
are lazy ... improved
motivation needed

supervisor has favourites
... supervisor allocates
tasks according to ability

better working relationships
... high staff turnover

Fig. 6.2   *Personal construct map of the 'workload' problem*

quences of both improved motivation and allocation of duties according to abilities.

The model is not an exercise in strict logic: the logical opposite of 'supervisor has favourites' is 'supervisor treats everyone the same' but this was not how the group saw the contrast. They chose instead to deal with a 'psychological' opposite which indicates how the group perceived favouritism as operating within their work domain. The group recognized that human relationships are inevitably affected by personal feelings and that it was not possible to screen out the effects of such feelings completely. Instead, they chose to make objective the aspect which they believed could be controlled and saw the 'psychological' opposite as 'supervisor allocates tasks according to ability'. This in turn revealed the possibility of a link to 'better working relationships' which reflected the group perception of what effect this change to practice could presage.

Rich pictures

This type of modelling causes, arguably, the greatest concern for those using soft systems methods for the first time. The problem is one almost of embarrassment at having to draw what appear to be cartoon characters whilst engaged in a serious discussion. If considerations of dignity can be set aside, however, the beginner will find that rich pictures are an effective way of capturing a set of perceptions and modelling the Worldview of individuals and groups.

A rich picture seeks to capture four aspects:[4]

- processes – who is doing what?
- structures – in what organizational framework?
- climate – under what cultural norms and values?
- issues – involving what problems and feelings?

In the course of a study each participant will need to talk about individual perceptions of these aspects. The analyst uses a rich picture to encapsulate these perceptions as a group statement about the opportunity or problem being investigated. The aspects listed above may be used by the analyst to provide a guide, but should not be used to control the discussion. Just as with cognitive mapping, the freedom of the technique allows perceptions to be included as they come to mind: the analyst should not force participants to 'exhaust' each category or aspect before moving on to the next. Although this may mean that the conversation is occasionally repetitive, or harks back to a previous point, such 'rambling' is often the best way to allow participants to develop confidence, talk about difficult issues, test ideas and reveal feelings.

Because rich pictures tend to be drawn freehand there are few rules and many individual styles have developed, reflecting both the drawing skills of the analysts, the problems being investigated and the comments made by participants. Drawing skills improve with practice but the use of a few simple conventions will assist the tyro and serve to promote confidence.

The aspect 'structures' is usually taken to connote things or concepts that are slow to change. Examples are the organizational roles and responsibilities captured in organograms, buildings, places and locations. These can be represented by icons of buildings, organograms and people. Another commonly used icon is of an eye, indicating external scrutiny or the need to meet standards. For example, employee unions might be monitoring the organization to assess the fairness of labour practices.

'Processes' represents an aspect which is subject to short-term and sometimes rapid change. Examples such as filing, travel, writing can be represented by icons such as a set of folders, an aeroplane and a pen. It is sometimes easier to attach to an icon a brief descriptive phrase or even to substitute the phrase for the icon. Individual judgement and dexterity are the best arbiters of the form to be used.

Expressing the aspect of 'climate' means linking the structures and processes in order to show how they interact and what external factors have to be taken into account. For example, linking the icon representing the organizational structure of the stock ordering department, described in the previous sections, with the icon representing the supervisor and the process of task allocation indicates an exercise which depends on the acceptance of legitimate authority as a power base.

The climate described above should be one of acceptance but the case study has already revealed that this is a process which has caused disquiet: this indicates that there is an 'issue' associated with the process. If the issue is contentious, this can be signalled by the use of crossed swords and emphasized by naming the point of conflict in a 'think bubble'. A rich picture based upon the example of the stock ordering department is shown in Figure 6.3.

Rich pictures tend to display processes, structures and issues, with climate being deduced from a consideration of the whole picture. In this respect it may be difficult for someone outside the group process to be able to read the picture and appreciate the relationships. This serves to emphasize the participative nature of the process of constructing a rich picture: it is simultaneously a record of thought and a means of generating, checking and extending that thought. Rich pictures can certainly be used as an aid to individual thought but their full value becomes apparent when working with a group.

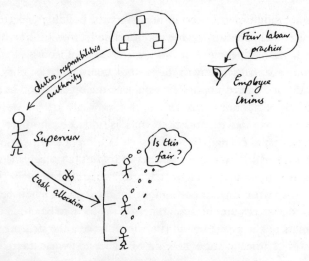

Fig. 6.3 *Rich picture of the 'workload' problem*

Documenting complex systems may necessitate the drawing of several pictures, each being a refinement of, or a component in, another picture. Keeping track of several levels of rich pictures can be quite taxing: the drawing of a 'master' picture showing each separate rich picture as an icon and recording the relationship of each with the others is helpful in keeping track of the place of a concept in the overall scheme.

A rich picture can be started at any point during an investigation, group meeting or other form of enquiry. As a first step it is frequently a good idea for the analyst to review the documentation already available about the organization and its activities and then summarize those structures, processes and issues which seem most pertinent to the problem or opportunity. This initial rich picture can then be used as the basis for discussion at the first meeting.

Using maps and rich pictures together

Some practitioners rely simply upon one of the modelling techniques described so far. There is something to be said for simplicity and becoming fully expert in using one approach, but in my experience there are occasions when another is needed, if only to break the monotony.

Using a personal construct map is, for example, an excellent way of exploring an issue which has been tentatively identified in a rich picture. A cognitive map forms a good resumé of a rambling discussion during which participants are settling down and talking about structure.

One of the key purposes of maps is to enable the analyst and the participants to gain an understanding of what issues are of significance in approaching the problem or opportunity. Patching[5] has identified two important clues to this process, 'fundamentality' and 'frequency'. It is problems which affect structure which are most obviously 'fundamental' because they are long-lasting, and large-scale changes may be necessary in order to rectify them. In addition, there are problems which are acute rather than chronic. Such problems are often in the forefront of the minds of participants because they relate to immediate issues and concerns. They can be identified by the frequency with which they are mentioned.

In noting such points there is an understandable tendency for the analyst and the group to regard fundamental problems as being of greater importance. This view may lead to the apparent sidelining of any point which is not regarded as fundamental: the analyst must be aware of this possibility and prevent the seemingly trivial from being forgotten unless it has been agreed to be of no further significance. Major improvements to levels of satisfaction in an organization can be achieved through such seemingly trivial exercises as an improved design of an official form. The key is to listen to the frequency with which such points are made: if the item is mentioned often then it is not one of which to lose sight.

Conceptual models

In Chapter 4 the notion of a conceptual model was introduced and it was observed to be an extension of the root definition. In particular it takes the Transformation, which has already been defined in the root definition, and prompts some consideration of the minimum activity that needs to be undertaken in order to achieve the identified Transformation. Checkland and Scholes[6] have emphasized the role of conceptual models as being a key part of debating how to achieve improvements in the 'real world' through changes which can be modelled in the 'systems world'. Conceptual models are consequently not representations of some aspect of the real world; they are logically related to root definitions developed in the systems world but with the aim of stimulating discussion about how a particular change can be achieved.

To be desirable, the rate and direction of a change must be capable of being monitored and controlled. Conceptual models must, therefore, include two types of activity:

- operational – those related to achieving the desired change
- supervisory – those related to monitoring and controlling the operations defined above.

and they must be clearly linked to the root definition.

Being concerned with activities, the terms used in building a conceptual model depend heavily on the use of verbs, these representing the minimum activities considered necessary to achieve the process of Transformation identified in the set of definitions produced by applying the CATWOE components.

The model must recognize logical and temporal dependencies. If part of the change is, for example, going to require the installation of a new photocopier, its purchase and the preparation of a suitable power supply must precede training in its operation. The ordering of suitable supplies of consumables are logically dependent on the choice of photocopier but are not temporally dependent on its installation; they must be available in time for the inception of its operation, however, so there is a temporal relationship involved with that action.

Conceptual models are intended to assist in planning and achieving the Transformation identified in the root definition. To be useful, a conceptual model should be confined to about seven to nine distinct operations: more will make the model too complex to hold easily in the mind and think about holistically. If, when developing a conceptual model, there seems to be a need to include more activities than nine, this is an indication that multi-level models are needed, with the primary level providing an overall model and serving as a key to lower-level models, each of which describes one of the activity points in the primary model. Each activity will then also need a root definition from which the lower-level conceptual model can be built.

The need to develop multi-level models also reinforces a point made in an earlier chapter: soft systems analysis is often iterative, prompting the analyst and participants to return to an earlier stage when it becomes apparent that a concept has not been completely or satisfactorily explored. Similarly, the first conceptual model which is built may not stand up to more rigorous examination when its workability is discussed. In particular, the criteria of the 'five Es' of efficacy, efficiency, effectiveness, ethicality and elegance (discussed in Chapter 4) may not be satisfactorily addressed by the actions specified in the model. Whether it is feasible, or worth re-entering the analysis because of a failure to satisfy one or more criteria will be a matter of debate but it is important that such matters are considered before the model is accepted.

Drawing a conceptual model

The actual drawing of the model can best be summarized in a general form suggested by Checkland;[7] this has been developed as Figure 6.4. This shows a conceptual model of a Transformation which has two inputs: some information and a triggering event. The receipt of information causes Action 1 to occur, whilst the occurrence of the trigger results in Action 2 taking place. Action 3 will not occur until both Action 1 and Action 2 have been completed. Action 3 is thus temporally dependent on Actions 1 and 2, though not logically dependent: it is possible for either Action 1 or Action 2 to occur first, or for both to occur simultaneously, but Action 3 will not occur until both have been carried out. Action 4 is also temporally dependent on Actions 3

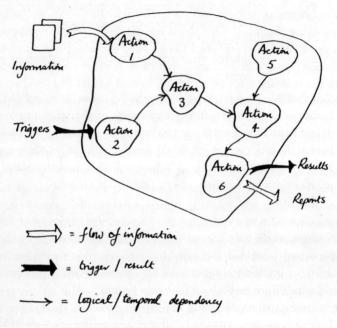

Fig. 6.4   *General form of a conceptual model*

and 5. Action 6 is logically dependent on Action 4: it cannot occur until Action 4 has been performed but depends on no other event. When performed it yields an output in two forms: a flow of information as a report and a result which may form the trigger to another conceptual model. The Actions are contained within a system boundary, serving to distinguish between the input to a system, the processes within and the output.

For example, in the case of the stock ordering department, if there has been a decision that task allocation is the key to the discontent and the process should be changed in order to reflect the abilities of the work group members, CATWOE components and the root definition of the desired system might be:

C	work group members
A	supervisor
T	allocation of tasks in stock ordering department

W	tasks should be allocated in accordance with the abilities of the members of the work group
O	Personnel Manager of the library service
E	employee unions regularly review the fairness of employment practices.

Root definition: A system, owned by the Personnel Manager of the library service, by which the supervisor of the stock ordering department allocates tasks according to the abilities of the members of the work group employed there. Employee unions monitor employment practices, including task allocation, to assess to what extent such practices meet a criterion of fairness.

During discussion of this definition it may become evident that 'fairness' means taking into account the skills of each member of staff, their existing workload and rate of work, and monitoring performance. This immediately implies a set of actions and a process of control which must be built into the new system.

The conceptual model for an improved system is shown in Figure 6.5. Information about tasks to be undertaken and deadlines are necessary information for the prioritization of tasks. Knowledge of the size of existing backlogs is the other input of information. When this information is combined with the results of a skill assessment of the work group members and an estimate of work rate it is possible to estimate the 'slack' in workload which each member of the group may have. Combining this information with that about task priorities then allows tasks to be scheduled. The final actions are to consult work group members about the proposed schedule and make any agreed adjustments. The output from the system is a task list showing allocations and deadlines for completion.

Considering this conceptual model against the criteria of the 'five Es' it may become evident that in one respect, at least, it is deficient: it is not efficacious in respect of the need to monitor progress, which is an important part of ensuring fair task allocation. However, trying to include this concept in the existing conceptual model will take it close to the recommended limit for complexity. A better approach will be to recognize that the action of 'monitoring performance' is

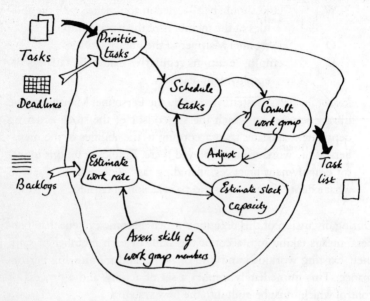

Fig. 6.5   *Conceptual model of the 'task allocation' system*

itself an action of some complexity which really needs a separate root definition and conceptual model. In turn, it is helpful to see the 'task allocation' conceptual model as part of a higher-order model, such as Figure 6.6, which shows feedback from the action 'Monitor performance' into the rest of the overall system. This level of model also shows the origin of the information about backlogs and tasks to be done, together with the deadlines. In other words, it identifies other parts of the job of the supervisor which will have to be defined and may require additional analysis.

As mentioned in Chapter 4, the conceptual model is the link between the systems world and the real world. Whilst in the systems world one does not screen out reality; however, the mental translocation allows the mind and creative instinct to think through how things ought to be before asking how that state can be achieved.

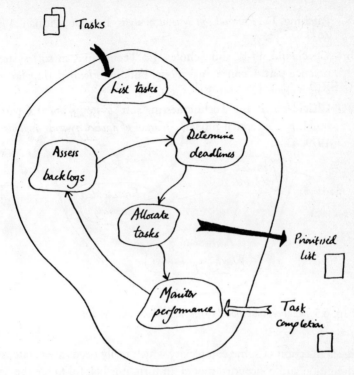

Fig. 6.6 *Conceptual model of the higher-order system for 'task allocation'*

References

1   Tomlinson, R., 'Of tools, methods, and methodology', in *Tackling strategic problems: the role of group decision support*, edited by C. Eden and J. Radford, London, Sage, 1990, 184.

2   Buzan, T. and Buzan, B., *Mind map book: radiant thinking – a major evolution in human thought*, rev. edn, London, BBC, 1995.

3   Eden, C., Jones, S. and Sims, D., *Messing about in problems: an informal structured approach to their identification and management*, Oxford, Pergamon, 1983, 39–59.

4   Woodburn, I., 'The teaching of soft systems thinking', *Journal of applied systems analysis*, **18**, 1991, 29.

5   Patching, D., *Practical soft systems analysis*, London, Pitman, 1990, 69.

6   Checkland, P. B. and Scholes, J., 'Techniques in soft systems practice part 4: conceptual model building', *Journal of applied systems analysis*, **17**, 1990, 39.

7   Checkland, P. B., 'Techniques in "soft" systems practice part 2: building conceptual models', *Journal of applied systems analysis*, **6**, 1979, 43.

Chapter 7

Defining functional requirements

THE DEVELOPMENT OF a conceptual model is the linking stage between the 'systems' and the 'real' worlds. Although it is not necessary to maintain a strict separation of the two, it is worth remembering that time spent in the systems world allows participants to consider what ought to happen to resolve a problem or make use of an opportunity before encountering whatever restrictions the real world may offer. This time is not wasted in idle speculation or 'stargazing': the development of improved services and products often depends on making creative proposals and then seeking the necessary resource base to bring them into reality.

On crossing the boundary back into the real world the conceptual model needs to be compared with reality in order to determine if it is feasible to undertake the intended change. Having considered the Human Activity system in some depth during the systems world phase of the study there should be few complications which arise from this comparison. Complications may however arise from lack of adequate equipment, materials, time or some other limited resource. In order to test feasibility, therefore, it is necessary to develop a set of functional requirements. This will in turn serve to define the resources required and indicate procedures which need to be planned or further defined before implementation can take place.

Hard systems analysis techniques, such as cost/benefit analysis, queuing theory and resource allocation modelling, may be needed to facilitate selection of suitable resources. The need to use another type of analysis is not because soft systems analysis has failed: hard systems analysis is especially suited to dealing with matters of choice where the objectives are clear. The exercise of soft systems analysis should have yielded such clarity.

Methods for developing functional requirements

Several 'packaged' methods have been developed which incorporate both soft and hard techniques of analysis. Patching[1] describes two which have been developed and widely used to assist in the implementation of computerized information technology. The packaged techniques are robust, in that they have been used successfully in many different circumstances, and flexible because they incorporate soft systems methods.

- *Functional Analysis of Office Requirements (FAOR)*[2] is especially appropriate for the design of information technology support for office work. Implicit in its use is that an office consists of a network of people engaged in structured, semi-structured and unstructured activities. The aim is firstly to assess the range of activities in an effort to understand the human activity system and the nature of the change which is required; soft systems analysis is an essential method for this aspect. The second step is concerned with considering and selecting suitable tools for the detailed analysis of what has been found in the human activity system. Use of either, or both, soft and hard tools may be found appropriate. The third step is to apply the tools in order to plan a schedule of the resources and procedures which need to be developed to achieve the desired change. This specification of requirements is then used for the detailed design and implementation of a system which will take account of these. These activities are outside of the scope of FAOR but the final step returns within its boundaries, when an evaluation of the operating characteristics of the new system against the requirements specification is undertaken.
- *Multiview*[3] is a technique aimed specifically at the problem of developing information systems which have to be technically efficient and an effective support for the people who use them or are affected by their use. There are five steps involved in using Multiview:

analysis of the set of human activities which will be the locus of operation of the information system; this information forms the background to the complete study

analysis of the data items (usually called 'entities'), processes and outcomes which constitute the information flow in the system; this information is passed to the final step

analysis and design of the socio-technic aspects of the system, paying particular regard to the choices and options available in order that an optimal solution can be recognized and adopted; this information is passed to the next step

human/computer interface design, in which decisions will be made about the procedures users will employ to work with the system and how the system will respond; the technical implications of this are passed to the next step

design of the technical aspects, which takes the information from the data item analysis and the technical implications of the interface design step and facilitates development of an efficient design and full requirements specification for the new system.

The introduction of information systems based upon computers is a common feature of work in many libraries, information services and resource centres and for these purposes one of the packaged methods just described could be very suitable. Both include the use of soft systems analysis so there should be an adequate recognition of the needs of the people involved and opportunities to refine objectives.

A general model for developing a functional requirement

Computers and the use of information technology are not the only areas which present problems and opportunities in the work of managing libraries, information services and resource centres. Within them, dealing with the Human Activity system forms a major part of the work of management and many projects may be undertaken which do not involve the use of computers or information technology except, perhaps, as an incidental or in a minor capacity. Such projects still need a means of transforming the conceptual model of the agreed and desirable change into a set of functional requirements. A paper by Hirscheim and Schafer[4] provides a basis on which such a general model can be developed. The model consists of four levels, each of which has a distinctive focus on which it is necessary for management to concentrate if a satisfactory system is to emerge.

Level	Focus
Resources	Efficiency and standards; lead time
Infrastructure	Channels and agents
Information flow	Linkages
Context	Quality

Each level is important for successful completion of a project but the nature of each project will determine whether every level is of similar weight or whether one or more levels will predominate.

Considering the Resources level, this encompasses the equipment, materials, buildings, services, people and skills which will be needed and the schedule for their use. By matching them against the resources immediately under the control of the management, a desiderata list and timetable for acquisition can be produced. For materials, equipment, buildings and services the focus for management in controlling the project is on the specifications that suppliers will have to meet in order for their products to match the demands of the new system. Many standards will be derived from the appropriate industrial code, specification or practice. In terms of staff the focus will be on determining what skills and personal characteristics are going to be needed in order to determine a personnel specification for people who might be recruited to new positions, and on the training that they, or staff already in post, might require in order to meet the required standards. All resources require a certain 'lead time' after ordering before they are actually available for use. This is also a critical factor for the project and constitutes part of the managerial focus at this level. An overall managerial focus for all types of resource is the efficiency of their use.

At the level of Infrastructure the requirements specification considers the stakeholders who are involved with the new system and with the system which it is replacing. Their relationship with the old and the new systems is noted because changes of any significance will need to be conveyed to them. The particular focus for management is therefore the channels for carrying such communication and the agents responsible for the messages which need to be sent.

The level of Information flow is concerned with considering the links which the existing system has to other processes. There are few systems which are independent and entire: most depend upon an input of information or trigger of some kind from another system and most also produce information or a trigger which is of significance to another system. The focus for management is to what extent the existing linkages are to be maintained. The new system must be capable of picking up the links if the network of systems is to continue to function. If the redesigned system cannot support all, or at least some of the links, then the process of design is incomplete and another round of study will be needed in order to determine what the effect of this might be.

Context is the final level. Here the new system is considered in relation to the rules, regulations and procedures which constitute the social and technical controls which surround it and to which its operation must adhere. The rules may be formal in the sense of having legal force, such as those to protect copyright and privacy of personal information; they may be informal, such as custom and practice. The managerial focus is on quality, assessing to what extent the new system meets the requirements established by the context.

An overall focus must be on effectiveness: the purpose of change and the design of a new system is to achieve greater effectiveness in satisfying user needs. The individual foci in the requirements specification examine particular aspects which have a bearing on effectiveness. In considering these foci, management will in turn be contributing to the attainment of improved effectiveness.

The role of project management

No requirements specification is complete without considering the need for project management. Project managers are people who are accountable, have legitimate power to use or divert resources and are cognisant of the overall timetable and individual milestones of project completion.

Depending on the scale of the project and the extent of the impending change, it may be desirable to make use of formal methods of project management which provide guidance on the scheduling and control of resources.

It is now time for the analyst to step back from the project. The analysis phase leads into the specification of requirements and signals a diminishing role for the analyst. The management of the implementation stage must be undertaken by the organization if it is to exercise a proper degree of ownership over the new system. To do otherwise is to pervert the relationship which should exist between the analyst and the rest of the people in the organization.

Good analysis is about illuminating a path, facilitating progress but rarely leading. Project management is quite definitely about leading and directing.

Conclusion

Throughout this book several case studies have been used to demonstrate some of the problems encountered in consultancy and how soft systems methods could have been used to deal with them. The value of case studies is the opportunity they provide to reflect upon how best to deal with problems, and to speculate on what the consequences of different approaches might be.

To remind you of a point made in the opening chapter, 'consultancy' is not an activity confined to consultants: it is a set of behaviours, techniques and attitudes aimed at clarifying the thoughts of others. In that sense, most effective managers use consultancy skills.

Improvement in the practice of consultancy depends on the sharing of experience, errors and successes. Soft systems methods can be used in many ways and, because they offer great freedom in their method of use, there is much to be gained from critical assessment and development of them in the analysis of problems and opportunities in libraries, information services and resource centres.

An old canard often attached to the practice of consultancy is that, for a large fee, a consultant will borrow your watch, look at it and then tell you the time. As with many such stories, it contains a degree of truth: but what needs to be added is that when the watch is returned, you know that it is both working and showing the correct time.

References

1   Patching, D., *Practical soft systems analysis*, London, Pitman, 1990. 135–8.

2   Schafer, G., *Functional analysis of office requirements: a multiperspective approach*, Chichester, Wiley, 1988.

3   Avison, D. E. and Wood-Harper, A. T., *Multiview: an exploration in information systems development*, Oxford, Blackwell, 1990.

4   Hirscheim, R. and Schafer, G., 'Requirements analysis: a new look at an old topic', *Journal of applied systems analysis*, **15**, 1988, 101–18.

Further reading

THIS READING LIST is grouped into a small set of categories structured around the main ideas and techniques introduced in this book. It is neither exhaustive nor are the categories necessarily mutually exclusive. This should not surprise the reader: soft systems analysis draws on many disciplines, the boundaries between which are often blurred. There are many books covering topics such as 'organizational behaviour': I have included only those which I have found useful, some of which are now out of print but still, in my view, worth reading.

Change management

Jones, K., *Conflict and change in library organizations: people, power and service*, London, Bingley, 1984.
A classic work which provides an excellent foundation for exploring organizational structure and the process of change. The early chapters explore classical models of structure such as bureaucracies, and move on to the human behavioural school before considering paths to organizational development.

Managing strategic action: mobilizing change concepts, readings and cases, edited by C. Hardy, London, Sage, 1994.
The main focus is on understanding the process of change and how it should be achieved. Several models of the relation between strategic choice and organizational structure are explored and there is an emphasis on the importance of understanding power in an organization. The examples are drawn from commerce and industry but many contain sufficient insights to be useful to the non-commercial library, information service and resource centre sector.

Underwood, P. G., *Managing change in libraries and information services: a systems approach*, London, Bingley, 1990.

A discussion of hard systems techniques, centring around the development and implementation of computer systems in libraries and information services.

Creative thinking and problem solving

Buzan, T. and Buzan, B., *Mind map book: radiant thinking – a major evolution in human thought*, rev. edn, London, BBC, 1995.

A challenging and stimulating text which presents many models of thinking with practical suggestions for releasing creativity. The use of 'mind maps' as an aid to documenting discussions and structuring thought is richly illustrated.

Eden, C., Jones, S. and Sims, D., *Messing about in problems: an informal structured approach to their identification and management*, Oxford, Pergamon, 1983.

The authors use their considerable experience of consultancy to discuss how to cope with problems which are inherently 'messy'. A case study, exploring marketing in the manufacturing industry, is used as a continuing theme through the book; the exploration of the case focuses on behaviour, however, and the context does not obscure the illustration of methods.

Flood, R. L. and Jackson, M. C., *Creative problem solving: total systems intervention*, Chichester, Wiley, 1991.

The 'total systems intervention' technique acknowledges the importance of the political dimension in dealing with problems in organizations. The framework for problem solving is firmly in the soft systems mould and also draws on associated techniques.

Tackling strategic problems: the role of group decision support, edited by C. Eden and J. Radford, London, Sage, 1990.

A valuable collection of papers from consultants which shows the variety of approaches yet also allows for some synthesis of techniques. The emphasis is on understanding context, assisting with analysis and the management of this process, rather than on prescribing solutions.

Weick, K. E., *Sensemaking in organizations*, London, Sage, 1995. (Foundations for organizational science).
A strong challenge to the notion that rationality is the only or even the best way to view an organization and its processes. In this account, realities are explored through creative reflection and comparison.

Organization theory

Abrahamsson, B., *The logic of organizations*, Newbury Park, Sage, 1993.
An excellent outline of classical theories of organization coupled with contemporary discussion on efficiency, the representativeness of organizational structures and how to overcome problems implicit in bureaucratic forms of organization.

Abrahamsson, B., *Why organizations?: how and why people organize*, Newbury Park, Sage, 1993.
A discussion of the conditions necessary for the emergence of hierarchies, bureaucracy and democracy in organizations.

Katz, D. and Kahn, R., *Social psychology of organizations*, 2nd rev. edn, New York, Wiley, 1978.
A foundation text for anyone working with organizational structures. Three main sections explore the theory of organizations, models and the problems and processes encountered in them.

Organizational behaviour

Bennett, R., *Organisational behaviour*, 2nd edn, Pitman, 1994. (M & E handbooks.)
An introductory text which provides an account of the classical, scientific management and 'human behavioural' models of behaviour.

Ivancevich, J. M. and Matteson, M. T., *Organizational behavior and management*, 3rd edn, Homewood, Irwin, 1993.
A thorough analysis of the field, with many examples and illustrations drawn largely from commerce and industry in the USA. The ideas can be considered independently of their context.

Nadler, D. A., Hackman, J. R. and Lawler, E. E., *Managing organizational behavior*, Boston, Little, Brown, 1979.
The use of a clear central model for exploring and understanding organizational behaviour makes this book especially useful. The authors emphasise that the book is not theory-bound but seeks to understand the types of behaviour likely to be encountered, their antecedents and possible outcomes.

Systems thinking and theory

Checkland, P. B., *Systems thinking, systems practice*, Chichester, Wiley, 1981.
The classic work, outlining the reasons for the emergence and growth of the soft systems movement, its relationship with other forms of action research and a methodological discussion.

Critical issues in systems theory and practice, edited by K. Ellis, A. Gregory, B. R. Mears-Young and G. Ragsdell, New York, Plenum, 1995.
A major set of papers documenting the growth of the systems movement and soft systems in particular. The contents include

discussions of theory, development of techniques and critiques of application.

Critical systems thinking: directed readings, edited by R. L. Flood and M. C. Jackson, Chichester, Wiley, 1991.
This set of papers includes many which explore the 'political' dimension of organizations and the development of techniques of analysis and intervention.

Flood, R. L. and Carson, E. R., *Dealing with complexity: an introduction to the theory and application of systems science*, New York, Plenum, 1988.
Complex systems are very commonly found: analysing them and representing their growth, stability and decay is often difficult. This book presents a solid framework for understanding how this can be achieved; some mathematical concepts are introduced.

Handbook of systems analysis: overview of uses, procedures, applications and practice, edited by H. J. Miser and E. S. Quade, Chichester, Wiley, 1985.

Handbook of systems analysis: craft issues and procedural choices, edited by H. J. Miser and E. S. Quade, Chichester, Wiley, 1988.
A third volume covering case studies is planned. These volumes provide a comprehensive study of hard analysis procedures, policy analysis and social aspects of systems work.

Wilson, B., *Systems, concepts, methodologies and applications*, 2nd edn, Chichester, Wiley, 1990.
An excellent summary of systems thinking which can serve as an introductory text or as a refresher.

Soft systems analysis

Avison, D. E. and Wood-Harper, A. T., *Multiview: an exploration in information systems development*, Oxford, Blackwell, 1990.
A detailed description of the Multiview technique, which incorporates many soft systems ideas. Several case studies, including some from non-commercial contexts, make this exposition especially helpful.

Checkland, P. B. and Scholes, J., *Soft systems methodology in action*, Chichester, Wiley, 1990.
A discussion of the development of soft systems thinking during the 1980s. Several case studies are included, one of which concerns the development of the Information and Library Services Department of Imperial Chemical Industries (ICI). This case study was previously published (Checkland, P. B., 'Achieving desirable and feasible change: an application of soft systems methodology', *Journal of the Operational Research Society*, **36**, 821–31) but is here presented with an updated critique.

Davies, L. and Ledington, P., *Information in action: soft systems methodology*, Basingstoke, Macmillan, 1991.
A useful introduction to and outline of soft systems approaches, with a focus on use rather than theory.

Patching, D., *Practical soft system analysis*, London, Pitman, 1990.
A practical guide to applying and using soft systems analysis. The theory is explained, techniques identified and then linked with theory through case studies. A method of using a database held on a computer is explained and illustrated.

Working with groups

Bertcher, H. J., *Group participation: techniques for leaders and members*, 2nd edn, Newbury Park, Sage, 1993. (Sage human services guides 10.)

An excellent practical guide of interest to all participants in group processes.

Bertcher, H. J. and Maple, F. F., *Creating groups*, Beverly Hills, Sage, 1977. (Sage human services guides 2.)
The key stages in the formation of a group, identification of roles and management of the process are clearly explained.

Journals

There are several journals which regularly or occasionally include papers on using soft systems methods. The 'nuclear zone' of productive journals is small, however, and assiduous searching of abstracts will reveal several papers describing applications of soft systems methods published elsewhere.

Information systems journal: an international journal promoting the study and practice of information systems, Blackwell Scientific, ISSN 1350–1917
This is a continuation of the *Journal of information systems*, which replaced the *Journal of applied systems analysis*.

Information and management: the international journal of information systems applications, Elsevier, ISSN 0378–7206.

Interfaces: an international journal of the Institute of Management Sciences and the Operations Research Society of America, Institute of Management Sciences, ISSN 0092–2102.

Journal of the Operational Research Society, Macmillan, ISSN 0160–5682.

Index

Conventions:
The alphabetic arrangement is letter-by-letter

Documents mentioned in the text are indexed under author and title; the page number containing the full citation has the abbreviation (cit.) attached.

Diagrams are shown by italic page numbers, e.g. *31*

irt = in relation to

The use of this abbreviation does not affect the filing order of the substantive element.

Soft Systems Analysis

and the management of libraries, information services and resource centres

Peter G. Underwood
Professor of Librarianship
University of Cape Towm

LIBRARY ASSOCIATION PUBLISHING
LONDON

Published by
Library Association Publishing
7 Ridgmount Street
London WC1E 7AE

First published 1996

British Library Cataloguing in Publication Data

A catalogue record for this book is available from the British Library

ISBN 1-85604-150-6

Typeset from author's disk in 10/13pt Aldine 401 by Library Association Publishing.
Printed and made in Great Britain by Bookcraft (Bath) Ltd, Midsomer Norton, Avon.